CRIMINOLOGICAL AND LEGAL CONSEQUENCES OF CLIMATE CHANGE

This edited collection, the result of an international seminar held at the International Institute for the Sociology of Law, Oñati, Spain in 2010, explores the potential legal and criminological consequences of climate change, both domestically and for the international community. A novel feature of the book is the consideration given to the potential synergies between the two disciplinary foci, thus to encourage among legal scholars and criminologists not only an analysis of the consequences of climate change from these perspectives but to bring these fields together to provide a unique, inter-disciplinary exploration of the ways in which climate change does, or could, impact on our societies. Such an inter-disciplinary approach is necessary given that climate change is a multifaceted phenomenon and one which is intimately linked across disciplines. To study this topic from the point of view of a single social science discipline restricts our understanding of the societal consequences of climate change. It is hoped that this edited collection will identify emerging areas of concern, illuminate areas for further research and, most of all, encourage future academic discussion on this most critical of issues.

Oñati International Series in Law and Society

A SERIES PUBLISHED FOR THE OÑATI INSTITUTE FOR THE SOCIOLOGY OF LAW

Recent titles in this series

The Legal Tender of Gender: Welfare Law and the Regulation of Women's Poverty
edited by Shelley Gavigan and Dorothy Chunn

Human Rights at Work
edited by Colin Fenwick and Tonia Novitz

Travels of the Criminal Question: Cultural Embeddedness and Diffusion
edited by Dario Melossi, Máximo Sozzo and Richard Sparks

Feminist Perspectives on Contemporary International Law: Between Resistance and Compliance?
edited by Sari Kouvo and Zoe Pearson

Challenging Gender Inequality in Tax Policy Making: Comparative Perspectives
edited by Kim Brooks, Åsa Gunnarson, Lisa Philipps and Maria Wersig

Emotions, Crime and Justice
edited by Susanne Karstedt, Ian Loader and Heather Strang

Mediation in Political Conflicts
Soft Power or Counter Culture?
edited by Jacques Faget

For the complete list of titles in this series, see 'Oñati International Series in Law and Society' link at www.hartpub.co.uk/books/series.asp

Criminological and Legal Consequences of Climate Change

Edited by

Stephen Farrall
Tawhida Ahmed
and
Duncan French

Oñati International Series in Law and Society

A SERIES PUBLISHED FOR THE OÑATI INSTITUTE FOR THE SOCIOLOGY OF LAW

·HART·
PUBLISHING
OXFORD AND PORTLAND OREGON
2012

Published in the United Kingdom by Hart Publishing Ltd
16C Worcester Place, Oxford, OX1 2JW
Telephone: +44 (0)1865 517530
Fax: +44 (0)1865 510710
E-mail: mail@hartpub.co.uk
Website: http://www.hartpub.co.uk

Published in North America (US and Canada) by
Hart Publishing
c/o International Specialized Book Services
920 NE 58th Avenue, Suite 300
Portland, OR 97213-3786
USA
Tel: +1 503 287 3093 or toll-free: (1) 800 944 6190
Fax: +1 503 280 8832
E-mail: orders@isbs.com
Website: http://www.isbs.com

British Library Cataloguing in Publication Data
Data Available

ISBN: 978-1-84946-186-3

Typeset by Compuscript Ltd, Shannon
Printed and bound in Great Britain by
CPI Group (UK) Ltd, Croydon, CR0 4YY

Acknowledgements

We would like to thank our funders for their generosity; the International Institute for the Sociology of Law (in Oñati, Spain) were kind enough to offer us a venue for our initial deliberations in April 2010, and we would like to thank Malen Gordoa Mendizabal and her staff for their excellent organisation of our event (and for dealing with volcanic ash as well as climate change). The travel to and from Oñati was supported by a grant from the British Academy and we thank them for their award. At Sheffield, we owe thanks to Sarah Beedham and Harriet Godfrey for booking flights and dealing with nervous, twitchy co-organisers stressing about Icelandic volcanoes. Rachel Turner at Hart did a superb job of handling the manuscript from initial proposal to completed book. Finally, we must thank our contributors and the others who attended the seminar in Oñati for their enthusiasm and for the quality of their thinking and writing.

Contents

List of Contributors

Dr Tawhida Ahmed is Lecturer in Law, School of Law, University of Reading.

Dr Peter Doran is Lecturer in Environmental Law and Sustainable Development, School of Law, Queen's University Belfast.

Stephen Farrall is Professor of Criminology, School of Law, University of Sheffield.

Duncan French is Professor of International Law and Head of the Law School, University of Lincoln.

Dr Matthew Hall is Senior Lecturer in Law and Criminal Justice, School of Law, University of Sheffield.

Mark Halsey is Professor of Criminal Justice, Law School, Flinders University, Australia.

Chris Hilson is Professor of Law and Head of the School of Law, University of Reading.

Dr Konstantina Kalogeropoulou is Senior Lecturer, Kingston Law School, Kingston University.

Dr Annamaria La Chimia is Lecturer in Law, University of Nottingham.

Dr Tom Obokata is Senior Lecturer at the Human Rights Centre, School of Law, Queen's University Belfast.

Catherine Redgwell is Professor of International Law, Faculty of Laws, University College London.

Nigel South is Professor of Sociology and Pro-Vice-Chancellor, University of Essex.

Sally Wheeler is Professor of Law, Business and Society, School of Law, Queen's University Belfast.

Rob White is Professor of Criminology, School of Sociology and Social Work, University of Tasmania, Australia.

Exploring the Legal and Criminological Consequences of Climate Change: An Introduction

BY STEPHEN FARRALL, TAWHIDA AHMED AND DUNCAN FRENCH

BACKGROUND

IT IS NOW beyond any reasonable argument that the consequences of climate change are both inevitable and likely to be considerable. Such consequences, although initially 'physical' in their nature (ie increases in temperature, floods, extreme climatic variations, soil erosion and desertification) are also going to be experienced along social and economic axes in the form of migration, losses or withdrawal of resources, displacement of peoples, cultural dislocation, anxiety, community stress and the such like. These processes will have knock on consequences for society and the legal and statutory systems that exist to regulate behaviour. While legal scholars have been considering the legal framework to tackle climate change for over two decades, both lawyers and criminologists have only recently started to think through, and holistically about, the consequences of climate change for European and global society.

It is entirely understandable that attention so far has been placed on the immediate physical ramifications of climate change (such as land erosion, increased flooding, habitat loss and threats to bio-diversity) and the immediate economic and social impacts of such changes (damage to buildings, loss of infrastructure, migration and ill health). However, consideration ought also to be given not only to the longer-term financial implications but also to some of the less tangible social, normative, legal and criminological impacts.

This book stems from a seminar (held in April 2010 at the International Institute for the Sociology of Law, Oñati, Spain, supported by a grant from the British Academy) which explored the likely legal and criminological consequences of climate change for the European Union and the international community, as well as potential synergies between the two disciplinary foci. Our aim in producing this collection is to encourage, amongst legal scholars and criminologists, a consideration of the consequences of climate change for these fields of research. The collection brings together scholars of criminology and law in order to provide a unique, inter-disciplinary exploration of the ways in which climate change does or could impact on our societies. Such an inter-disciplinary approach is necessary given that climate change is a multifaceted phenomenon and one which is intimately linked across disciplines. To study this topic from the point of view of a single discipline would restrict our understanding of the concerns of climate change. Our ultimate objective is to identify emerging areas of concern; suggest directions forward; and to illuminate urgent areas for future research.

OUR FOCUS AND THE THEMES OF THE COLLECTION

We start with two chapters aimed at setting the scene for much of what follows. Farrall explores what climate change might do to our societies and hence to our experiences of crime and the consequences of climate change unfold in time and space. His 'thinking through' of what might happen (pieced together from what we know about climate change and the causes of crime) stands as an invitation to others to explore in greater detail how such processes may actually unfold (or are unfold*ing*). For example, increases in the variability of weather systems, and the attendant problems of forecasting 'weather events' (such as the storms/floods and bush fires in Europe in the summer of 2007) will pose problems for those estimating insurance risks. In light of this, insurers may increase premiums 'across the board', creating an impression that they are exploiting the situation in order to increase profits. This may lead to an erosion of confidence in the insurance sector, and lead to widespread mis- or over-claiming amongst consumers as they try to 'fight back' against what they see as exploitative prices. As increased demands are placed upon member state and EU-provided services by climate change, so taxation (both of citizens and businesses) may need to be increased in order to pay for such services. Since many people may not wish to pay increased rates of taxation, so rates of taxation avoidance may increase. This will have consequences for not just those services provided for by taxation, but will also see increased demands placed upon those who regulate and police the taxation system. Redgwell examines the international legal challenges of a lower carbon future, in particular whether existing international legal tools are 'fit for the purpose'. This includes analysis not only of the existing climate change legal regime but also of alternatives to carbon based fuels and the legal implications of 'technological fixes' such as carbon capture and storage and geo-engineering.

Hilson provides an intriguing account of how climate change litigation is evolving, from a social movement perspective. Hilson argues that framing (the struggle over ideas and the meanings attached to them) has been central to many 'green' social movements' actions and reactions to these. Thus, the police, prosecution authorities and media reporting of climate change protest have tended to apply a 'hard' frame, whereby the protesters are constructed as at best criminals or, at worst, extremists or terrorists who are threatening not only law and order but also industrial and aviation security. The climate change movement has subsequently attempted to reframe using 'soft' counter-frames, in which they seek to construct themselves as 'reasonable' citizens, acting in the public interest, and upholding the rule of law.

White explores how particular strategies adopted as solutions to the problem of climate change are, in turn, generating new forms of social and environmental harm. Examples of such paradoxical harm include the adoption of compact fluorescent light globes to save energy (but which contain toxic mercury) and the promotion of nuclear energy (but which involves disposal of nuclear waste). Bio-security in the context of climate change, as informed by notions of the national interest, is also linked to harms associated with climate-induced migration, conflicts over food and water, and the transformation of food crops into bio-fuels. From the point of view of an eco-justice perspective, the chapter questions those strategies designed to mitigate or adapt to the effects of global warming in ways that create different or additional kinds of social and ecologic harm. Wheeler explores the extent to which shareholder activism and the creativity captured by intellectual property rights that exist within the boundaries of the corporation can be harnessed to make concern for the environment and future enjoyment of the environment central to business activity.

Several chapters in the book focus on climate change and security, human rights and migration. Thus, South explores the challenges that climate change brings to rights, civil society movements and the research agenda in the twenty-first Century and Obokata complements this chapter by considering whether a human rights approach can, and should be, applied to climate change. La Chimia investigates the use of development aid assistance by developed countries to address climate change issues and to protect the environment and in so doing explores whether rights within developing countries are enhanced or confounded by the tying of development aid to environment protection goals. Kalogeropoulou examines possible consequences for supplementary pension rights of workers who move between EU member states and the issues that arise in relation to the preservation of relevant pension entitlements. All of these chapters push forward the idea that our conception of terms like 'justice' and 'human rights' may well face challenges as the realities of living with climate change become ever more pressing and apparent.

One of the recent features of international capital has been the tendency for large firms to move capital and resources around the globe relatively easily and with little or no concern for the economic and social impacts of this (eg the opening of overseas call centres, the location of industrial plants in countries with lower staff costs, etc). If, as seems likely, certain parts of the globe start to appear to be geographically or geo-politically insecure, so firms will move capital and resources from such places to safer locales. This 'migration' of capital will leave some areas with little or no legitimate forms of employment, possibly furthering forms of illegal activity. Halsey thus examines the impact of regulation of forests on capitalism, carbon sinks and climate change.

Climate change may force us to start to reconsider some basic ideas of what it means to be a member of society. For example, how will climate change affect our understanding of democracy?; do issues like climate change provide a much more direct connection between the electorate (thus furthering grass-roots democratic movements) or is the threat of climate change better handled at the technocratic level and removed from popular decision-making? At a more global level climate change is by definition a global issue; will it help consolidate the idea of (green) global citizenry, or will it simply promote isolationism and protectionism? Moreover, do issues like climate change alter the nature of the interface between global order and its inception into national consciousness and national regulatory structures, encouraging more permeable legal delimitations? And in addressing climate change, one must be aware that it has been incorporated within the wider debate on 'security'; to what extent should environmental issues of this magnitude be considered alongside and as significant as other concerns, currently of more immediate pre-occupation (most notably terrorism)? Doran's chapter speaks to such concerns via a consideration of personal carbon trading schemes, as does Hall's on the relationship between the State and the likely victims of climate change by promoting an interdisciplinary dialogue between victimology and international environmental law. Doran draws on Nikolas Rose's reading of Foucault for insights into the modern construction of 'citizen' and 'consumer'. In particular, Doran relies on Rose's discussion of the new specification of the subject of government to propose that the concept of Personal Carbon Trading (PCT) sits at the apex of a number of debates of interest to the environmental policy community (This included climate change, sustainable consumption and individual freedom in an era of global environmental and resource constraints).

Ahmed and French challenge the 'multilateral' paradigm as the singular approach to climate change in the international law regime. They consider three pairs of competing narratives ('analytical <-> normative', 'atomistic <-> integrationist' and 'localised <-> globalised') which are useful in enhancing our understanding of climate change and environmental law as moving beyond multilateralism. These 'pairs' are not mutually exclusive, demonstrating that international legal responses to climate change cannot be viewed through uni-focal lenses. The narratives also amplify the tensions which exist within and across narratives, bringing into sharp relief both the need to embrace the diversity of international legal approaches to climate change, while at the same time ensuring that the capacity to use climate change for each and every approach does not gravely endanger the consensus and drive needed to secure its meaningful progress.

Increased immigration to Europe and other regions, such as North America and Australasia is a distinct possibility as some countries in Africa and Asia become climatically less and less habitable. As well as leading to potentially large numbers of displaced peoples, such climatic changes could lead to tensions within affected countries leading to secondary waves of war-induced migration. Will the current levels of human rights protection provided to these different identity groups cope in the new climate? Will public perception exacerbate the problems, leading to restrictive political will to accommodate the human rights concerns of these new communities? Indeed, will all of this lead to a retreat on the willingness to continue to provide rights to existing communities? There may be an increase in racism and xenophobia. As well as increased immigration to Europe, they may well be increased migration within Europe, partly as a 'knock-on' effect, but also as a result of some parts of Europe becoming uninhabitable or less desirable. Accordingly a number of our contributors (principally Farrall, South, Kalogeropoulou and Hall) touch on some of the issues surrounding migration and reflect on what this could mean for legal and criminal justice systems and frameworks.

This collection represents some of the initial forays into this area of research by criminologists and legal scholars. What is clear is that both—alongside sociologists, human geographers, those with an interest in cultural change, psychologists and those studying health—will need to get to grips with these and related issues as the consequences of climate change start to become ever more apparent. We hope that this collection can both assist in helping us grapple with climate change and stimulate other criminologists and legal scholars to pursue these issues as part of their own research agendas.

1

Where Might *We Be Headed? Some of the Possible Consequences of Climate Change for the Criminological Research Agenda*

STEPHEN FARRALL*

These findings from Sand Canyon Pueblo suggest that climate-induced food stress and consequent violent conflict contributed to the depopulation of the Mesa Verde region in the late A.D. 1200s.[1]

INTRODUCTION

IF THE RECENT reports on climate change are to be believed, and there is growing evidence that they are to be,[2] then the impacts of climate change on the Earth's eco-system are going to be both dramatic and far-reaching. While it is entirely understandable that government and scientific attention be placed on the immediate physical ramifications of these changes (such as coastal erosion, flooding and habitat loss) and the immediate economic and social impacts of such changes (such as damage to buildings or the degradation or total loss of infrastructure) consideration ought also to be given, not only to the longer-term financial implications,[3] but also to some of the less immediately obvious impacts.

* I would like to express my thanks to the following people for comments on earlier drafts of this chapter and for helping me to track down various publications: David Gadd; Clare Jones; Constance Lever-Tracy and my co-editors.

[1] KA Kuckleman, 'The Depopulation of the Sand Canyon Pueblo' (2010) 75 (3) *American Antiquity* 497.

[2] IPCC (Intergovernmental Panel on Climate Change), *The Physical Science Basis, Contribution of Working Group I to the 4th Assessment Report to the IPCC* (Cambridge, Cambridge University Press, 2007); O Boiral, 'Global Warming: Should Companies Adopt a Proactive Strategy?' (2006) 39 (3) *Long Range Planning* 315–30.

[3] N Stern, *Review Report on the Economics of Climate Change* (London, HM Treasury, 2006).

authors suggest that 'there is no foreseeable political or technological solution that will enable us to avert many of the climatic impacts projected here', p55). This scenario predicts an average global temperature rise of just 1.3°C above the 1990 levels by 2040, rising to 2.8°C above 1990 levels by the end of the twenty-first century.[9] However, while the average increase in temperatures is 'just' 1.3°C, this equates to an increase of 2°C over land (where it is generally warmer than over large bodies of water). This is likely to lead to an increase in the levels of glacial ice melt, leading in turn to a rise in sea levels by about a fifth to a quarter of a metre (by 2040). They suggest that:

> Global mean sea level increases by 0.23 meters, causing damage to the most vulnerable coastal wetlands with associated negative impacts on local fisheries, seawater intrusion into groundwater supplies in low-lying coastal areas and small islands, and elevated storm surge and tsunami heights, damaging unprotected coastlines. Many of the affected areas have large, vulnerable populations requiring international assistance to cope with or escape the effects of sea level rise. Marine fisheries and agricultural zones shift poleward in response to warming, in some cases moving across international boundaries.[10]

And go on to add that:

> Regionally, the most significant climate impacts occur in the southwestern United States, Central America, sub-Saharan Africa, the Mediterranean region, the mega-deltas of South and East Asia, the tropical Andes, and small tropical islands of the Pacific and Indian Oceans. The largest and most widespread impacts relate to reductions in water availability and increases in the intensity and frequency of extreme weather events. The Mediterranean region, sub-Saharan Africa, northern Mexico, and the southwestern United States experience more frequent and longer-lasting drought and associated extreme heat events, in addition to forest loss from increased insect damage and wildfires. Overall, northern mid-latitudes see a mix of benefits and damages. Benefits include reduced cost of winter heating, decreased mortality and injury from cold exposure, and increased agricultural and forest productivity in wetter regions because of longer growing seasons, CO_2 fertilization, and fewer freezes. Negative consequences include higher cost of summer cooling, more heavy rainfall events, more heat-related death and illness, and more intense storms with associated flooding, wind damage, and loss of life, property, and infrastructure.[11]

If this all sounds pretty nasty, one has to bear in mind that the next scenario ('severe') is premised on the notion that many of the assumptions underlying the IPCC's report are unduly optimistic (in that they ignore feedback

[9] However, even these projections may be beyond our reach; see, 'Worst Ever Climate Emissions Leave Climate on the Brink' *The Guardian*, 29 May 2011. Alarming news, especially since the global recession was hoped to reduce emissions.

[10] Campbell *et al*, above n 8, 41.

[11] Campbell *et al*, above n 8, 41.

processes), and that the average global temperature rise might be nearer to 2.6°C above 1990 levels, and with a sea level rise of 0.52 metres within 30 years from 2007—ie by 2037. Campbell *et al* suggest[12] that the severe scenario will result in the following: reductions in water availability; reductions in crop yields; and lower levels of yields from ocean fisheries. The 'catastrophic' scenario (based on a temperature rise of nearly 6°C and two metres of sea-level rise) is truly terrifying. However, being based on the likely situation 100 years hence is probably too far off for any degree of certainty to be attached to it—other than, that is, that the future looks extremely bleak.

Dyer[13] suggests that the severe scenario is likely to lead to several of the world's largest cities being inundated due to rising sea-levels and the continual establishment of new ports and abandonment of these as the sea encroaches ever upwards. Crop yields will decline dramatically as usable land is lost to the sea or the costs of irrigating it become too great. Populations which cling to the coasts in some parts of the world will be forced inland or will start to try to move to parts of the world which are less prone to flooding or drought. It is to a consideration of this that I now turn.

HOW WILL THIS START TO IMPACT UPON AND SHAPE OUR SOCIETIES? HOW WILL SUCH RESHAPING AFFECT INDIVIDUAL HUMAN AND SOCIETAL EXPERIENCES? HOW WILL THESE VARY ACROSS COUNTRIES, REGIONS AND BY THOSE KEY SOCIO-DEMOGRAPHIC VARIABLES WE KNOW TO BE ASSOCIATED WITH LIFE CHANGES AND LONG TERM OUTCOMES?

It is to a consideration of this that I now turn. What are the secondary consequences of this for people/populations? What will this do to human societies?

Before we go on to look at what the likely consequences of the above may be for crime, we need to examine what *may* happen to our societies generally, since this will provide us with some clues as to what may happen to crime. Again my focus is largely at the macro-level, focusing on whole systems impacts, rather than the immediate consequences for individuals.

Lever-Tracy,[14] in her own consideration of what climate change may do to the existing varieties of capitalism, suggests that the pressing need to tackle climate change may challenge the basis of our modern economic institutions and those with an interest in maintaining them.[15] The political, economic and social ideologies which foster and which are fostered by current forms of capitalism may face some challenges as a result of some of the

[12] Campbell *et al*, ibid 71–74.

[13] G Dyer, *Climate Wars* (Oxford, Oneworld Publications, 2010) 15–23.

[14] C Lever-Tracy, 'Alternative Scenarios: Varieties of Capitalism' in C Lever-Tracy (ed), *The Routledge Handbook of Climate Change and Society* (London, Routledge, 2010) 274–79.

[15] Lever-Tracy, ibid 274.

consequences of climate change. Capitalism, however, is likely to endure as a general approach to the economy; the changes in it may, however, bring new winners and new losers, which may bring to the fore a new set of relationships between government and the industrial sectors. For instance, if the 'small is beautiful, local is best' ideology firmly takes hold, then small businesses serving local needs may profit, while larger entities may face a withdrawal of public support.

However, the most telling part of Lever-Tracy's essay concerns human values and how these may be affected by climate change. As she notes, values which emphasise the short term and the immediate may cease to appeal to many people. The emphasis on immediate profits and super-individualisation are relatively recent features of society, and values which promote respect for the environment and a longer term, less individualistic stance may come to the fore. As evidence of this, Lever-Tracy points to the fact that many migrants take jobs in societies which do not value their real skills or qualifications so that their children may have a better life and also that the tsunami of 2004 saw a global outpouring of sympathy and concern for those effected—even though they be socially, culturally and geographically distant strangers. As Tracy-Lever writes 'values of discipline, solidarity and loyalty to family and community can become elevated over those of individual satisfactions and freedoms'[16] during periods when core social values change dramatically and speedily. These periods of 'intergenerational cultural landslides'—where the core values of one's parent's generation are overturned—may be provoked by dramatic changes (such as climate change).[17] If more and more tightly-focused regulation is required in order for businesses to show that they are complying with any emerging green agenda, so neo-liberalism, which has tended to view regulation with askance, may itself face ideological challenges (see Doran, this collection). Of course, the above reading may prove to be too optimistic; as nations are forced into accepting those fleeing from countries which are no longer viable, so we may witness an increase in the distinctions between 'them' and 'us', with all of its attendant xenophobia, intolerance and hatred. Such developments, which may be sharply experienced in some locales but not others, may result in the general breakdown of society and/or the creation of two parallel societies—one of 'pre-mass migration residents' and one of 'mass-migrants'.

One of the recent features of international capital has been the tendency for large firms to move capital and resources around the globe relatively easily and with little or no concern for the economic and social impacts of this

[16] Lever-Tracy, ibid 276.

[17] Evidence of such value change may be found in the UK in the sudden shift away from the emotional and institutional support for a state-organised economy geared (broadly) towards the production of an egalitarian society and the resulting shift (from the early 1970s) towards desires to 'own one's own home', 'work for oneself', 'get ahead in life' and the acceptability of income differentials (aka 'neo-liberalism').

(witness the opening of overseas call centres, the location of industrial plants in countries with lower staff-costs or weaker health and safety regimes, etc). If, as seems likely, certain parts of the globe start to appear to be geographically or geo-politically insecure, such firms may move capital and resources from such places to safer locales. This 'migration' of capital will—again—leave some areas with little or no legitimate forms of employment. Such possibilities make migration—either for economic reasons or for the simple need to find somewhere to live—all the more likely. We might then expect to see a decrease in the population in some areas of the globe and an attendant increase in some other parts of the globe. Campbell *et al* suggest that for Europe

> Environmental pressures will accentuate the migration of peoples to levels that effectively change the ethnic signatures of major states and regions. In Europe the influx of illegal immigrants from Northern Africa and other parts of the continent will accelerate and become impossible to stop, except by means approximating blockade. There will be political tipping points marked by the collapse of liberal concepts of openness, in the face of public demands for action to stem the tide. As the pressure increases, efforts to integrate Muslim communities into the European mainstream will collapse and extreme division will become the norm.[18]

Accordingly, some communities—the wealthy most obviously—may try to insulate themselves from such 'invaders', perhaps consolidating differences between the 'haves' and the 'have-nots'. Citing Leon Fürth, Dyer argues that religious and ideological turmoil will emerge, along with authoritarian political rule.[19] The Campbell *et al* report suggests that some coastal areas of the USA may start to become uninsurable[20]—along similar lines to the ways in which some parts of New Orleans became uninsurable after Hurricane Katrina in 2005. Increases in the variability of weather systems, and the attendant problems of forecasting 'weather events' (such as the storms/floods and bush fires in Europe in the summer of 2007) will certainly pose problems for those estimating insurance risks in the developed world. In light of this and the likelihood of increased rates of loss, insurers may increase premiums for many common forms of insurance. Such policies may result in some people finding themselves uninsurable or needing to pay more for their insurance. This, in turn, may lead to an erosion of confidence in the insurance sector amongst consumers.

In some cases, whole business models may shift away from their present locations. For example, Hanson and colleagues[21] suggest that increases in droughts, fire risk, water shortages, heat stress, risk of tropical diseases and loss of habitat pose considerable risks for the viability of the current

[18] Campbell *et al*, above n 8, 74.

[19] Dyer, above n 13, 22.

[20] Campbell *et al*, above n 8, 72.

[21] CE Hanson, JP Palutikof, and MTJ Livermore, *et al*, 'Modelling the Impact of Climate Extremes: An Overview of the MICE Project' (2007) 81 *Climate Change* 163–77, 174.

Mediterranean tourist industry (which contributes significantly to the economies of Greece, Spain and more recently Portugal[22]). Furthermore, they suggest that the above factors, with an attendant increase in the Northern European summers, would lead to a decline in numbers of people wishing to travel to the Mediterranean for holidays.

Climate change looks likely to affect poverty levels,[23] with rural communities and people living in poorer parts of the globe bearing the brunt of these changes. Large-scale reductions in rural wealth, as well as leading to further out-migration, also threatens to damage the sustainability of rural communities in many countries initially affected by climate change (most likely lying on the circumference of the globe). There may be competition between communities over access to vital resources (such as clean drinking water and food—not to mention land itself). Wealthy nations appear already to be purchasing large tracts of land in developing parts of the world (most notably in Africa[24]) as an insurance against not being able to provide enough for their own citizens from within their own borders.

The costs of providing the food required by some nations may well see increases in price (due to scarcity and increases in transportation costs). Witness attempts by Algeria, Indonesia, Saudi Arabia, Libya and Bangladeshi in early 2011 to buy up surplus grain, even though these were at inflated prices.[25] As food prices rise, so some nations may need, at least initially, to contribute towards the costs of food. Social security systems may need, initially at least, to provide food vouchers for more vulnerable members of society, for example. Increased migration may require (at a statutory level) the 'receiving' nation to provide housing—again at a cost to the tax payer. If the recent history of migration to many European nations is any indication of what lies in store, migrants can expect to live, in most cases, in the most basic of accommodation and for it to be (or to eventually become) among the least desirable sections of our cities. Such processes of segregation along ethnic and geographic axes will serve only to reinforce the distinctions between the haves and the have-nots, and may lead to further examples of 'gated communities' and/or enclaves of the wealthy or the poor. If migration outstrips what many societies can provide easily, then 'shanty towns' on the edges of major European cities may become a real possibility.[26]

[22] Tourism represents about 15% of Greece's GDP, 11% of Spain's and 5% of Portugal's.

[23] R Mendelsohn, A Basist, P Kurukulasuriya and A Dinar, 'Climate and rural income' (2007) 81 (1) *Climatic Change* 101–18.

[24] Dyer, above n 13, 101.

[25] The Daily Telegraph, 'Egypt and Tunisia Usher in the New Era of Global Food Revolutions' (30 January 2011).

[26] The Guardian (16 May 2011, page 21) suggests that the Mongolian capital (Ulan Bator) now has a shanty town of some 700 000 (of a total population of 2.7 m). The existence of this shanty town is attributed to desertification caused by global warming and over grazing of cattle.

Employment may also be hard for migrant groups to attain—again in common with past recent experiences, whereby some migrant groups experience lower rates of engagement in the economy (Bangladeshis in the UK, for example, have some of the highest rates of unemployment). The reasons for this may be many and varied: lack of skills; inability to demonstrate skills to the levels required by western employers; insufficient need for those skills held by migrants and so on. Common employment practices in many developed nations (where pre-employment checks, CVs and skill portfolios are becoming increasingly common for secure employment) may serve only to exclude many from gaining legal work—not to mention language or other, more formal, barriers to employment. Whatever may happen, as increased demands are placed upon European member states and EU-provided services by climate change, so taxation (both of citizens and businesses) may need to be increased in order to pay for such services. Similarly, basic goods and services may also experience dramatic and sustained price inflation as the costs of procuring such goods increase and suppliers pass on costs to the consumer and/or suppliers exploit basic needs (bearing comparison with Thompson's notion of the moral economy). Although no such plans have been unveiled, it may be that access to some goods becomes limited (eg dwindling fish stocks[27]), either through the operation of normal market practices (ie price inflation) or through direct government intervention in the market place.

Given that access to food, water and basic security is now considered to be a key part of the provisions of the liberal state (even if, of late, some of these services have been contracted out to private companies) their withdrawal or rises in their scarcity and hence costs may trigger something of a crisis of confidence in the state. As Campbell *et al* note:

> One way or the other, severe climate change will weaken the capacity of liberal democratic systems to maintain public confidence. This intensified search for spiritual meaning will be all the more poignant under conditions of severe climate change. Governments with resources will be forced to engage in long, nightmarish episodes of triage: deciding what and who can be salvaged from engulfment by a disordered environment. The choices will need to be made primarily among the poorest, not just abroad but at home. We have already previewed the images, in the course of the organizational and spiritual unravelling that was Hurricane Katrina. At progressively more extreme levels, the decisions will be increasingly harsh: morally agonizing to those who must make and execute them—but in the end, morally deadening.[28]

[27] E Stenevik and S Sundby, 'Impacts of Climate Change on Commercial Fish Stocks in Norwegian Waters' (2007) 31 (1) *Marine Policy* 19–31.

[28] Campbell *et al*, above n 8, 73.

They also hypothesise what may happen to the *wider* political basis of, in this case, the USA:

> The United States' federal system may also experience stress. As noted above, one possible consequence of severe climate change will be greatly increased frequency of region-wide disasters as the result of an increasing number of especially violent storms. At some level, even a well-prepared Federal Emergency Management Agency (FEMA) system might be overwhelmed. As the cumulative magnitude of such damage increases, the federal government would likely leave state governments to shoulder more and more of the burden. The effect would be to strain the ligaments that hold the federal system together.[29]

They go on to add that 'At some point the government's ability to plan and act proactively will break down because the scale of events begins to overwhelm policies before they can generate appreciable results'.[30] Writing about Europe, Campbell *et al* suggest that

> At severe levels of climate change, civil disorder may lead to the suspension of normal legal procedures and rights. The precedents for dealing with large, unwanted minorities have already been set in Eurasia under fascism and communism. Under conditions marked by high levels of civil confusion and fear, political leaders and movements will emerge who might not resist these solutions.[31]

These scenarios—all likely within the next 30 years or so—raise a number of pressing and troublesome questions. How will the state respond to threats which are not from other states (aka 'war'), or small groups of people which threaten it (aka 'terrorists') or a larger but more diffuse and socially differentiated group of people for whom it must care and in some way respond to (aka 'the elderly', 'the criminal', 'the unemployed', etc) but from non-human, non-state threats (ie climate change)? These questions—as pressing as they are—are not my concern herein; my focus now is on what the above sorts of impacts on human societies are likely to do to our experiences of crime, policing and security.

HOW WILL THESE SECONDARY CONSEQUENCES FIND EXPRESSION AS CRIME?

It is obvious that climate change will not affect all places in the same ways, at the same time or speed. Most of the materials relating to those regions most likely to be affected first suggest that it is those regions of the world around the equator which will be affected first (in terms of temperature rises, etc) with some parts of the globe initially witnessing only mild rises in temperatures. Asia, Africa and South America are likely to experience these effects

[29] Campbell *et al*, above n 8, 73.
[30] Campbell *et al*, above n 8, 73.
[31] Campbell *et al*, above n 8, 74.

first.[32] Let us try to think through how the criminological consequences of climate change will play out in something approaching a sequential order.

First, then, let us examine what may happen in those societies *directly affected* by climate change (by which I mean those countries nearest to the equator). Campbell *et al* suggest some very unpleasant outcomes:

> Severe climate change will likely be the deathblow for democratic government throughout Latin America, as impoverishment spirals downward. In these circumstances we should expect that populist, Chavez-like governments will proliferate. Some regions will fall entirely and overtly under the control of drug cartels. Some governments will exist only nominally, and large regions will be essentially lawless, much as has been the case in Colombia. The United States will lack adequate means for responding effectively, and will likely fall back on a combination of policies that add up to quarantine.[33]

Similarly, British sociologist John Urry suggests that one of the possible long term outcomes for societies is what he refers to as 'regional war-lordism',[34] characterised by oil and gas wars, the breakdown of systems of communication, a decline in standards of living, the control of recycled weapons and vehicles by feudal chiefs, all with minimal forms of governance or regulation. So, according to Urry and Campbell *et al*, some countries will fall into the hands of violent extremists or organised criminals. But on the way to this, there will be other, perhaps more mundane consequences which are played out in what is, at least initially, a set of circumstances very similar to what exists at present. In short, we will not go from industrialised, 'minority-world' societies to *Mad Max*[35] overnight.

Large-scale reductions in rural wealth, as well as leading to further im/migration will threaten to damage the viability of rural communities and may possibly lead to increases in both unemployment and consequently crime in such communities. As food becomes scarce so there may first be price hikes (leading perhaps to localised riots about food prices, as happened in the UK during the eighteenth century)[36] and then larger scale shortages. Such episodes may lead people to turn to illegal markets for basic food stuffs. Undoubtedly such illegal markets will be run by those who will be willing to use threats or actual violence to ensure that they get the price they want for the goods they sell, and who may also be involved in other allied trades (such as the supply of weapons, drugs and the control of prostitution). Perhaps—for all of this remains in the realms of speculation

[32] Dyer, above n 13, 4.

[33] Campbell *et al*, above n 8, 73.

[34] J Urry, 'Climate Change, Travel and Complex Futures' (2008) 59 (2) *British Journal of Sociology* 261–79, 268–69.

[35] *Mad Max* (1979); an Australian film directed by George Miller, written by George Miller and Byron Kennedy and starring Mel Gibson, which tells the tale of societal breakdown in a dystopian future in which law and order are minimal.

[36] EP Thompson, *Customs in Common* (London, Penguin, 1991) 189.

of course—some of those young men without work in such regions will become involved in these businesses and perhaps there will be competition between rival groupings for the rights to provide services, etc (which may in turn lead to 'turf wars'). If this sounds overly pessimistic, bear in mind that the 'migration' of capital away from insecure regions or countries will leave some areas with little or no legitimate forms of employment. If the situation remains unchecked—and in some cases it might well, as governments discover that there are 'no go areas' in some countries for them, or if climate change forces a concentration on other matters (such as negotiations over how and where to evacuate people)—such rivalries may escalate into full-blown wars. Although it would be a gross over-simplification to attribute the war in the Darfur region of Sudan to climate change alone (since other factors were at operation too), the droughts of the 1980–84 seasons certainly played their part in this tragedy. With wars or large-scale on-going turf wars one can expect to see rises in crimes such as illegal arms dealing; rape; torture; murder and other violent crimes. However, many of these activities will—while being crimes—not make their way into courts or to those systems of crime-recording. Few people will attempt to take, for example, providers of illegally-sourced food to court (since they need the illegal markets themselves) and few policing systems will be able to cope with the levels of offending involved during times of extreme distress (in some cases the police may actually be part of a system of corruption anyway). So, we can expect that there will be a lot of crime, but little impact on crime statistics (where these are collated anyway).

Such events, as they escalate, would inevitably lead to the destabilisation of governments. However, even without armed conflict some crimes may become more common; extortion; people smuggling and allied trades may become more frequent. If people regularly go hungry they may take to the streets and forms of protest may become increasingly tense and violent. Again, although not the sole cause of the unrest, inflation in food prices were one of the reasons cited for the protests in Tunisia in late 2010 and early 2011. In the case of Tunisia the protests became so widespread and intense that it led to the collapse of the government. As well as leading to potentially large numbers of displaced peoples, such climatic changes could result in tensions within affected countries leading to secondary waves of war-induced migration.[37] And herein lies another lesson; climate change may lie behind some of the popular uprisings of the future, but the role of climate change may be as a *contributing* factor to—rather than a *sole cause of*—any such protests.

Eventually, as has often been the case in the past, refugees will start to leave their once-homelands (either as a direct result of climate change making some

[37] N Brenjo, 2007, 'Looking to Water to Find Peace in Darfur' in *Reuters Alert*: www.alertnet.org/db/blogs/1265/2007/06/30-100806-1.htm (accessed 15 October 2007).

of their land unviable or as a result of persecution or the fear of persecution). What will happen—in terms of crime—in those countries to which migrants flee? This brings me to what will happen to those countries which are (at least for a short while) likely to be *indirectly* effected by climate change.

Even ignoring, at least for a moment, the impact of migration upon various countries (Europe, North American and the more stable states in Africa, such as South Africa), and to which I shall turn to presently, there are a number of possibly criminogenic consequences of climate change. As noted above, increases in the variability of weather systems, and the attendant problems of forecasting such 'weather events' will pose problems for the estimation of insurance risks. Whatever they do, insurers will need to increase at least some premiums.[38] If such rises reach a level at which a widespread feeling that insurers are exploiting the situation in order to increase profits, we may see an increase in attempts to defraud insurers.

This may lead to an erosion of confidence in the insurance sector, and lead to widespread mis- or over-claiming amongst consumers. In 2004 Karstedt and Farrall[39] asked citizens in England and Wales about the extent to which they trusted their insurers to make them a fair offer during a claim; around 20 per cent said that they did not. They also asked people if they had ever been offered less by their insurer than they felt that were entitled to; 27 per cent said that this had happened to them at least once. When asked if they were worried about being left out of pocket on an insurance claim some 39 per cent said that they were very or fairly worried about this possibility. Five per cent said that they had deliberately cheated on an insurance claim. Imagine what may happen to such figures if insurers start to withdraw coverage from some parts of the country or if they start to raise their premiums. Insurance may seem like an odd service to highlight in a discussion of the consequences of climate change. However, for many people who see themselves as 'non-criminal', insurance claims offer one arena in which they could extract (an illegal) profit without feeling that they were offending. As insurers become more picky about who or what they are prepared to cover and how much of an excess they are prepared to request, so insurance clients may seek to gain redress via such scams.

There are also issues as to how climate change might affect our approach to regulation; will there be a continued move away from command-and-control techniques towards more market-oriented mechanisms or, alternatively—once it is found that the latter does not alter personal behaviour sufficiently—will there be a return to more coercive regulatory measures? Along these lines, governments may increasingly become required to monitor and reduce waste products and/or to avoid landfill, with

[38] Leichenko *et al*, above n 4, 143.

[39] S Karstedt and S Farrall, 'The Moral Economy of Everyday Crime: Markets, Consumers and Citizens' (2006) 46 (6) *British Journal of Criminology* 1011–36.

such as the English Defence League. Recent qualitative research conducted from a psycho-social framework has suggested that

> White people's feelings of relative disadvantage led us to conclude that part of the appeal of racism lay in its capacity to act as a receptacle for many unacknowledged emotions: shame about their inability to secure decent lives for themselves and their families; anxieties about ill-health and the risk of criminal victimisation; and humiliation as their investments in the locality [Stoke on Trent, England] were exposed as unsophisticated or imprudent.[46]

As such, the losses (both perceived and real, both subjective and material) experienced by the individuals that Gadd and Dixon spoke to were a key aspect of their attraction to racist discourses (and for some, activities). Reading the above, it is not hard to see how the long term marginalised sections of the White 'working class', given both the past, historic losses (jobs, community, roles) and some of their likely future, climate-change-induced losses may find the overtures stemming from organisations such as the British National Party or the English Defence League hard to resist. This may have further ramifications for both actual assaults on migrants and on the acceptability to others of racially-tinged punitive sentiments.

The impacts of migration at the European end could therefore range from increased community tension, rises in 'people trafficking', a rise in anxiety amongst European citizenry,[47] which may express itself as 'fear of crime',[48] eroding trust in the criminal justice system and increasing pressures for punitive sentences, and hence, possibly, rises in rates of imprisonment for crimes relating to trafficking and subsistence (prostitution, drug-importation/selling, etc) to abuses of human rights. If those people fleeing loss of habitat or threat of harm elsewhere are socially and geographically excluded from our towns and cities (either deliberately or because the institutions and organisations of civil society are unable to cope with the levels and speed of arrivals) so we may witness an increase in 'shanty towns' on the edge of European cities (or at key border points). In a European social climate which is increasingly seen to be suffering from clashes between different religious, national and ethnic groups, the steady addition of diverse communities from Asia and Africa could see further strain placed on the already tense issue of multiculturalism versus assimilation in Europe. Will the current levels of human rights protection provided to these different identity groups cope in the new climate?

[46] D Gadd and B Dixon, *Losing the Race* (London, Karnac, 2011) 12.

[47] K Norgaard, 'People Want to Protect Themselves a Little Bit: Emotions, Denial and Social Movement Nonparticipation' (2006) 76 (3) *Sociological Inquiry* 372–96.

[48] S Farrall, J Jackson and E Gray, 2006, 'Everyday Emotion and the Fear of Crime: Preliminary Findings From Experience and Expression', Working Paper No 1, ESRC Grant RES 000231108.

Will public perception exacerbate the problems, leading to restrictive political will to accommodate the human rights concerns of these new communities? Indeed, will all of this lead to a retreat from the willingness to continue to provide rights to existing communities? There may be an increase in racism and xenophobia. Linguistically also, issues arise in relation to the balance to be struck between the provision of minority linguistic rights and the requirement for individuals to participate in the new societies they are joining. All of these concerns are underpinned by sensitivities surrounding the preservation of cultures and identities. As well as increased immigration to Europe, there may well be increased migration within Europe, partly as a 'knock-on' effect, but also as a result of some parts of Europe becoming uninhabitable or less desirable. The incidence of 'people trafficking' is already a high concern for European countries. How will the UK/Europe respond to this? Might this create a further pull in the direction of stringent collaborative action? What are the implications of this for the human rights of those trafficked—many of whom end up working in illegal and poorly regulated sectors of the economy?

Another casualty of the changes likely to be wrought by climate change may be our outlook on the future. It is human nature to want to see improvements in living conditions and lifestyles, both over one's own lifespan and for future generations. This may provoke an attitude shift towards 'greener' values, or, more pessimistically, it may provoke a loss of hope for the future. The ability to hope that 'things will be better in the future' has been linked to changes in attitudes towards crime at the individual level (see Farrall and Calverley)[49] and can be seen in the work of others such as Green[50] (in discussing the treatment of child murders) and Karstedt[51] (in discussing wider political cultures and the role these played in the rehabilitation of ex-Nazis) when exploring national-level responses to crime. A decline in levels of hope may bring with it an increase in levels of cynicism and scepticism—neither of which sound good from the point of view of the ideals associated with the rule of law. Finally, some crimes have been demonstrated to be influenced by seasonality/temperature.[52] As climatic shifts alter local temperatures, so human behaviour will change, with people spending (perhaps) longer away from

[49] S Farrall and A Calverley, *Understanding Desistance from Crime, Crime and Justice Series*, Chapter 5 (London, Open University Press, 2006).

[50] D Green, 'Political Culture and Incentives to Penal Populism' in H Kury (ed), *Fear of Crime—Punitivity: New Developments in Theory and Research* (Bochum, Universitatsverlag Brockmeyer, 2008) 251–76.

[51] S Karstedt, 'Life After Punishment for Nazi War Criminals: Reputation, Careers and Normative Climate in Post-War Germany' in S Farrall *et al* (eds), *Escape Routes: Contemporary Perspectives on Life After Punishment* (London, Routledge, 2011) 240–75.

[52] S Field, *Trends in Crime and Their Interpretation: A Study of Recorded Crime in Post-War England and Wales, Home Office Research Study* 119 (London, HMSO, 1990).

their homes (increased risk of burglary?) or (perhaps) more time at home (increased risk of domestic violence?). Field found that 'good weather' (ie a lack of rain and the presence of sunshine) was associated with increases in sexual offences, but decreases in non-residential burglary and shop theft.[53] Anderson[54] found that heat increases were associated with increases in violent motivations and behaviours, as did Rotton and Cohn.[55]

However one approaches the issue of crime, given the likely changes in the global climate over the next 30 to 50 years, it is hard not to imagine a situation in which there will be an increased focus on crime (and associated topics such as policing, regulation, justice and hostile sentiments)—even if in some cases crime is suppressed or there are no major increases in crime. Given this, it seems prudent to end this foray into the possible consequences of climate change for crime with a consideration of how well-placed criminology is to deal with these likely changes.

HOW WELL IS CRIMINOLOGY PLACED TO RESPOND TO SUCH EVENTS? WHICH THEORIES MIGHT ASSIST US IN FORECASTING AND PLANNING?

By way of a conclusion I want to spend some time considering how well placed criminology is and criminologists are to assist in what can be called the global 'coming to terms with' climate change and what it will entail for our societies and our experiences of crime within those societies. Given the uncertainties I have outlined above, this section is hugely underdeveloped.

There has been little research in the UK, for example on the impact of ethnic minorities on experiences of crime and rates of offending. The evidence that does exist (which is from the US) suggests that ethnic minorities tend to reduce local area crime rates. However, such studies have a number of problems. It is hard to fathom exactly how large number of ethnic minorities suppress crime, and therefore it is hard to know if such a general relationship will hold when one has to migrate whole communities or societies. Perhaps the US has benefitted from selection effects in migration more than we are aware. In any case, the US has traditionally been the melting pot of the world, and strong national values of individualism may act in such a way as to enable the presence of ethnic minority groups to suppress crime. On a related topic, the group threat theory (see Chircos *et al*[56]) provides a way of unpicking some of the sensibilities towards

[53] Field, ibid 50.

[54] C Anderson, 'Heat and Violence' (2001) 10 (1) *Current Directions in Psychological Science* 33–38.

[55] J Rotton and E Cohn, 'Global Warming and US Crime Rates' (2003) 35 *Environment and Behavior* 802–25.

[56] Chircos *et al*, above n 43.

ethnic minority groups and how these may unfold over time and are linked to objective changes in levels of migration and so on. However, at present there is no data set on this with a time series component (although such a data set could easily be incorporated into existing surveys—both related to crime and other social issues).

If people's engagement in crime is related to their relationship to key institutions and organisations (see Messner and Rosenfeld, 2001,[57] Sampson and Laub, 1993)[58] then, with those societies in possible states of strain (or at very least a state of flux and uncertainty) further work on the impact of climate change on actual offending rates may do well to borrow insights from Institutional Anomie Theory. In *Crime and the American Dream*, Messner and Rosenfeld extended the scope of anomie theory (Merton, 1938),[59] arguing that all social institutions are affected by structural and cultural imbalances in the economy. From this perspective, institutional anomie arises when the economic sphere is disembedded from other social institutions and when the culture and values of markets dominate other sectors and vital institutions of society, like the family, education and welfare. If the economy is itself likely to undergo dramatic shifts and challenges as a result of some of the consequences of climate change, then we may see a need to continually examine the relationship between shifts in the economic base and the consequences of this for feelings of anomie and attachment to social institutions.

At an individual level, it is likely that future generations of humans (regardless of whether they have had to endure long and pain migrations or have had the relative luxury of remaining in one place) will suffer significant losses. Such losses may be in terms of region and place of residence; homes; possessions; family members; careers; assumptive worlds and future; identities, etc, etc. In this respect, the recent work of Gadd and Dixon (2010)[60] and of Kauffman (2002)[61] and Farrall (2009)[62] may provide insights into what we might expect at the individual level. Such losses may, if they become common enough, transcend the individual level and form the basis of social movements (the English Defence League being a regressive example of this, the green movement being a more progressive example)—what these hold for crime itself will vary, of course, but my point is that the

[57] SF Messner and R Rosenfeld, *Crime and the American Dream*, 3rd edn (London, Wadsworth Publishing, 2001).

[58] R Sampson and J Laub, *Crime in the Making* (Cambridge, MA, Harvard University Press, 1993).

[59] RK Merton, 'Social Structure and Anomie' (1938) 3 *American Sociological Review* 672–82.

[60] Gadd and Dixon, above n 46.

[61] J Kauffman, *Loss of the Assumptive World* (London, Brunner-Routledge, 2002).

[62] S Farrall, 'We Just Live Day-to-Day: A Case Study of Life After Release Following Wrongful Conviction' in R Lippens and D Crewe (eds), *Existentialist Criminology* (London, Routledge, 2009) 169–96.

awareness of such losses and the importance of them to our understanding of what these will do to our experiences of crime are ones with which we need to become better acquainted.

CONCLUSIONS

What I have tried to do above is to 'think through' some of what may happen to societies' experiences of crime as the consequences of climate change start to become increasingly apparent and pressing. Doubtless I have missed vital clues as to what may happen, and may have wrongly interpreted some of the clues I have detected. However, what is clear to me is that there is a need for criminologists to start to think about what may happen and how we, as a community of scholars working in what is an essentially empirically-grounded area of research, may best respond to these challenges.

2

International Legal Responses to the Challenges of a Lower Carbon Future: Energy Law for the Twenty-first Century

CATHERINE REDGWELL

INTRODUCTION

WHATEVER VIEW ONE takes of the 'peak oil' debate,[1] there is no doubt that petroleum is a finite resource and that at some stage in the twenty-first century dependence on other forms of energy will surpass petroleum as the principal fuel source for developed economies. Indeed, a recent study[2] identifies increased energy demand and consequent depletion of convention fossil fuels as one of five factors driving the transition to a lower carbon future, namely: (i) climate change and other environmental damage resulting from fossil fuel use; (ii) energy security,[3] requiring diversity as well as security of supply; (iii) increased demand for energy (with consequent resource depletion, energy security and climate change impacts): (iv) market, political and institutional restrictions on conventional energy investment;[4] and (v) the need to eradicate energy poverty

[1] Namely, that physical sources of oil and gas and other carbon-based fuels have, or will, peak and become exhausted: see further M Simmons, *Twilight in the Desert, the Coming Saudi Oil Shock and the World Economy* (New Jersey, John Wiley and Sons, 2005) and P Roberts, *The End of Oil: On the Edge of a Perilous New World* (New York, Mariner Books, 2004).

[2] DN Zillman, C Redgwell, Y Omorogbe and L Barrera-Hernandez (eds), *Beyond the Carbon Economy: Energy Law in Transition* (Oxford, Oxford University Press, 2008).

[3] See further B Barton, C Redgwell, A Ronne and DN Zillman (eds), *Energy Security; Managing Risk in a Dynamic Legal and Regulatory Environment* (Oxford, Oxford University Press, 2006).

[4] Under its 'business as usual' reference scenario, the International Energy Agency (IEA) estimates that US$20 trillion will be required to be invested in energy facilities to meet conventional energy demand, over half of this sum in electricity generation and transmission where there is a major investment gap in the developing world: IEA, *World Energy Outlook 2009* (Paris, OECD/IEA, 2009).

Sources of International Energy Law

In the energy field, treaties are the principal source of legal rules. However, there is no global treaty instrument regulating 'energy' as such: indeed, the only energy treaty not limited to a particular source of energy or type of energy activity is the 1994 Energy Charter Treaty, a regional east–west European initiative which does not enjoy global participation.[15] Instead, the global picture is a fragmented one, with treaties addressed to a variety of transboundary impacts[16] and activities where international cooperation is required for energy activities to proceed.[17] Thus we find extensive regulation of the impact of the maritime transport of oil,[18] of oil pollution and nuclear liability and compensation,[19] of the safety of international shipping, offshore installations and nuclear installations,[20] and protection of the marine environment.[21] Cross-border energy investment is protected by bilateral and regional instruments where foreign property is protected, inter alia, from expropriation without compensation.[22] Rules have developed

[15] The Energy Charter Treaty (ECT and Protocol on Energy Efficiency and Related Environmental Aspects (PEEREA available at www.encharter.org. The preamble to the ECT explicitly recalls, inter alia, the UNFCCC and the 1979 ECE Convention on Long-Range Transboundary Air Pollution. It has 51 States party, plus the EC and Euratom, and number of observer States (eg the United States, China, and several OPEC countries). Russia is not a party but applied it provisionally until 18 October 2009.

[16] Eg the regional ECE 1991 Espoo Convention on Environmental Impact Assessment in a Transboundary Context and 2003 Kiev Protocol on Strategic Environmental Assessment, available at www.unece.org/env/eia. Annex I of the Convention requires prior environmental assessment of a number of energy activities such as crude oil refineries, thermal power stations, and large-diameter petroleum pipelines.

[17] Eg joint development agreements for the exploitation of fields straddling international boundaries. For recent treatment see P Cameron, 'The Rules of Engagement: Developing Cross-Border Petroleum Deposits in the North Sea and Caribbean' (2006) 55 *International and Comparative Law Quarterly* 559.

[18] Eg the 1973 International Convention for the Prevention of Pollution from Ships, as modified by a 1978 Protocol (1973/78 MARPOL) available at www.imo.org.

[19] Eg the 1969 International Convention on Civil Liability for Oil Pollution Damage and 1971 International Convention on the Establishment of a Fund for Compensation for Oil Pollution Damage, as amended, available at www.imo.org; the regional 1960 Paris Convention on Third Party Liability in the Field of Nuclear Energy and the 1963 Brussels Supplementary Convention, as amended, available at www.unece.org; and the global 1963 Vienna Convention on Civil Liability for Nuclear Damage, as amended, and 1988 Joint Protocol linking the regional and global nuclear liability regimes, available at www.iaea.org.

[20] Eg the 1988 IMO Convention for the Suppression of Unlawful Acts Against the Safety of Maritime Navigation extended to fixed platforms on the continental shelf by its 1988 Protocol, available at www.imo.org; and the 1994 Nuclear Safety Convention, available at www.iaea.org.

[21] There are a considerable number of international treaties addressed to marine pollution, including from energy sources additional to oil transport by sea, at the global and regional levels. See further the discussion in P Birnie, A Boyle and C Redgwell, *International Law and the Environment*, 3rd edn (Oxford, Oxford University Press, 2009) ch 7.

[22] In addition to over 2000 bilateral (general) investment treaties now in existence, energy investment is addressed regionally by the 1994 ECT and under chapter 11 of the 1992

for the settlement of energy disputes not just between States, but between investors and host States.[23] Energy trade and cross-border provision of energy supply and services are also increasingly regulated.[24] Of course, this fragmented picture is by no means unique to international energy law and can be seen as a by-product of the horizontal and decentralised character of the international legal system itself.[25]

In terms of substantive rules of customary international law of particular importance to the energy sector, three in particular may be noted: permanent sovereignty over natural resources;[26] the obligation not to cause environmental harm to the territory of other States or to areas beyond national jurisdiction;[27] and the duty to notify and cooperate with

North American Free Trade Agreement. For recent treatment see P Cameron, *International Energy Investment Law: the Pursuit of Stability* (Oxford, Oxford University Press, (2010) ch 4.

[23] Eg the ECT relies on three methods for dispute settlement between investors and host States by arbitration pursuant to ICSID, Stockholm Chamber of Commerce, or UNCITRAL rules.

[24] Eg additional to regional rules, both the WTO GATT and GATS are of general application to energy transit, trade and services, and there are an increasing number of project-specific treaty and contractual arrangements pertaining to transboundary energy infrastructure projects. See, for example, discussion of the Baku–Tblisi–Ceyhan, Nord Stream, and Chad–Cameroon pipelines, in C Redgwell, 'Contractual and Treaty Arrangements Supporting Large Transboundary Pipeline Projects: Can Adequate Human Rights and Environmental Protection be Secured?' and M A Bekhechi, 'The Chad–Cameroon Pipeline: Lessons Learned from the Legal Challenges Faced and Addressed by World Bank Financing' in M Roggenkamp, Lila Barrera-Hernandez, Donald N Zillman and Inigo del Guayo (eds), *Understanding the Legal Aspects of Energy Networks* (Oxford, Oxford University Press, in press).

[25] See, generally, the Report of the Study Group of the International Law Commission on the Fragmentation of International Law: Difficulties arising from the Diversification and Expansion of International Law A/CN4. /L.682, 13 April 2006, available at www.ilc.org, and G Hafner, 'Pros and Cons Ensuing from the Fragmentation of International Law' (2004) 25 *Michigan Journal of International Law* 849.

[26] In its judgment in *Case Concerning Armed Activities on the Territory of the Congo (Democratic Republic of the Congo v Uganda)*, (2005) ICJ, Rep para 22 *et seq*, the ICJ reiterated that the principle of permanent sovereignty over natural resources (PSNR) is part of customary international law. See generally N Schrijver, *Sovereignty Over Natural Resources: Balancing Rights and Duties* (Cambridge, Cambridge University Press, 1997); on the relationship between this principle and domestic concepts of property, see C Redgwell, 'Property Law Sources and Analogies in International Law' in A McHarg, B Barton, A Bradbrook and L Godden (eds), *Property and Law in Energy and Natural Resources* (Oxford, Oxford University Press, 2010) and R Barnes, *Property Rights and Natural Resources* (Oxford, Hart Publishing, 2008).

[27] The first contentious case before the ICJ after World War II, the *Corfu Channel* Case (1949) ICJ Reports 1, established the principle that States may not permit their territory to be used in such a fashion as to cause harm to other States. The 'no harm' principle was also famously elaborated upon the interstate arbitration between the United States and Canada in *Trail Smelter Arbitration* (1941) 35 AJIL 684 and has been reiterated in many instruments subsequently, including the Principle 21 of the non-binding Stockholm Declaration on the Human Environment and Principle 2 of the non-binding 1992 Rio Declaration on Environment and Development. It has clearly passed into customary international law, and was recognised as a general principle of international law in the *Case Concerning the Gabcikovo/Nagymaros Project (Hungary v Slovakia)* (1997) ICJ, Rep 7.

other States regarding risks arising from hazardous activities, including accidents.[28] There is also a duty to cooperate over shared natural resources in international law, which has developed principally in the context of shared water resources, to preserve a balance of rights between upstream and downstream riparian users, and over fixed bodies of water.[29] There is a procedural obligation to cooperate with respect to straddling petroleum fields—to notify, inform and consult, and to negotiate in good faith—but no substantive requirement in terms of outcome: there is no customary international law obligation requiring States to agree jointly to develop straddling hydrocarbons.[30] Other principles have not achieved independent customary international law status,[31] though they may bind as a treaty obligation if embedded in a treaty text, and may exert considerable influence over the interpretation of existing rules of international law.[32] Such principles, concepts or approaches include the principle or concept of sustainable development;[33] the precautionary principle or approach;[34] the polluter-pays principle;[35] and intergenerational equity.[36]

[28] This duty to notify is evident in both hard and soft law: see further P Sands, *Principles of International Environmental Law*, 2nd edn (Cambridge, Cambridge University Press, 2003) 456 and Birnie, Boyle and Redgwell, above n 21, 177.

[29] See the *Lac Lanoux* Arbitration (1957) 24 *International Law Reports* 101 and G Handl, 'The Principle of "Equitable Use" as Applied to Internationally Shared Resources' (1978) 14 *Revue Belge de Droit International* 40.

[30] Cameron, above n 147.

[31] For general discussion see Sands, above n 28 and Birnie, Boyle and Redgwell, above n 21 and N de Sadeleer, *From Policy to Principle*: Emerging Norms of International Environmental Law (Oxford, Oxford University Press, 2002).

[32] For excellent analysis of the principle of sustainable development in this regard, see AV Lowe, 'Sustainable Development and Unsustainable Arguments' in AE Boyle and D Freestone (eds), *International Law and Sustainable Development* (Oxford, Oxford University Press, 1999).

[33] The *concept* of sustainable development was recognised by the ICJ in *Gabcikovo* (n 27) and in the 'Legality of the Use by a State of Nuclear Weapons in Armed Conflict Advisory Opinion' (1996) *ICJ Rep* 1, although it stopped short of recognition of the principle of sustainable development as a matter of customary international law. Equivocal treaty support is found in Article 19 ECT. The literature on this topic is vast: for recent general treatment of the principle see D French, *International Law and Policy of Sustainable Development* (Manchester, Manchester University Press, 2005); for the legal status of its buttressing principles, see P Sands, 'International Law in the Field of Sustainable Development' (1994) 65 *British Year Book of International Law* 303. On specific linkages between PSNR and sustainable development, see Schrijver's 1997 analysis, above n 26, and R Lyster and A Bradbrook, *Energy Law and the Environment* (Cambridge, Cambridge University Press, 2006) ch 3.

[34] For analysis of the varying definitions of precaution and its consequent lack of customary law status, see Birnie, Boyle and Redgwell, above n 21, 159–64.

[35] First articulated in a 1973 OECD nonbinding instrument on 'The Polluter Pays Principle', the absence of consistent state practice, including its equivocal articulation in soft law instruments (eg 'the polluter should, in principle, pay' in Principle 22 of the nonbinding Rio Declaration and similarly in Art 19 of the binding Energy Charter Treaty) further undermines its potential as a customary international law norm.

[36] While concern for future generations has been expressed in the non-binding preamble to a number of treaty instruments (and, exceptionally, in the text of the 1992 Framework Convention on Climate Change, Art 3), and has been noted in the judgments of the ICJ, most particularly in the dissenting judgments of Judge Weeramantry, the principle of

There is another important category of norms of pertinence for international energy activities, referred to as soft law.[37] This consists primarily of declarations and recommendations made by international conferences, institutions, and organisations intended to influence the conduct of States (and non-State actors) but for the breach of which the traditional enforcement mechanisms of international law will not apply. A good example is the 1989 IMO Guidelines and Standards for the Removal of Offshore Installations and Structures on the Continental Shelf.[38] Since codes of conduct may well form the basis of national legislation, and/or be transformed into binding international law through treaty negotiation or into customary international law as a result of State practice (that is to say, 'hardened'), the role of codes of conduct and guidelines as contributors to the international law-making process is potentially a highly significant one. Such codes and guidelines may have a particularly important role to play in consensus-building with respect to the regulation of new technologies, such as geoengineering,[39] and/or permit those most directly affected (eg scientists involved in geoengineering R&D, concerned citizens) the opportunity directly to participate in norm creation. That said, it is important to realise that the transition from 'soft to hard' is not an ineluctable one, and that instruments of putative soft law character may remain in the shadowy territory of non-law. Certainly the attempts to generate norms of conduct for multinational enterprises inhabit this territory.[40]

A good illustration of the interplay between hard and soft law sources of international law is the principle of permanent sovereignty over natural resources, the bedrock of international law regulating energy resources

intergenerational equity is not recognised at customary international law: see French, above n 32 and C Redgwell, *Intergenerational Equity and Environmental Protection* (Manchester, Manchester University Press, 1998).

[37] See further C Chinkin, 'The Challenge of Soft Law' (1989) 38 *International and Comparative Law Quarterly* 850; D Shelton (ed), *Commitment and Compliance: the Role of Non-Binding Norms in the International Legal System* (Oxford, Oxford University Press, 2000); and C Redgwell, 'International Soft Law and Globalization' in B Barton *et al* (eds), *Regulating Energy and Natural Resources* (Oxford, Oxford University Press, 2006).

[38] The 1948 IMO Convention (289 UNTS 48) entered into force on 17 March 1958 and currently has 169 parties representing about 98% of world gross registered tonnage. Note that the legal effect of the Guidelines is enhanced by Art 60(3) of the 1982 United Nations Convention on the Law of the Sea (LOSC) which requires States to take into account any generally accepted international standards established by the competent international organisation, which clearly encompasses the 1989 IMO Guidelines. Thus, while the Guidelines are not directly binding as such, the States Parties to the 1982 LOSC have this weak treaty obligation to 'take them into account'.

[39] Such soft law guidance may emerge from the current initiative by the Royal Society, the Environmental Defense Fund, and TWAS, the Academy of Sciences for the developing world, on solar radiation management governance: see further www.royalsociety.org/SRM-governance-initiative.

[40] See, generally, P Muchlinski, *Multinational Enterprises and International Law*, 2nd edn (Oxford, Oxford University Press, 2007).

Institutions

In the light of the absence of a single global treaty addressed to energy as a whole, it is unsurprising to note that there is likewise no single global institution exclusively responsible for energy activities. Indeed, with regard to organisations active in the energy sector, it is possible to distinguish between organisations which have an energy mandate only (eg the International Atomic Energy Agency (IAEA), the IEA, and the International Renewable Energy Agency (IRENA)), those which have comprehensive mandates which enable them by way of interpretation to deal with energy (eg the UN), and other special organisations whose mandate may allow for particular energy activities as a means to achieve the organisation's aims (eg the International Maritime Organization (IMO)). There are also regional organisations with either a single energy mandate or a more general one allowing for energy activities, eg the Organization of Petroleum Exporting Countries (OPEC) and the Organization of Latin American Countries (OLADE). What is distinctly absent from this international landscape is a global organisation of general scope with open membership exclusively concerned with energy activities; there is no 'world energy organization' nor is there the political appetite amongst States to create an energy entity analogous to, say, the WTO.[54] International energy law in the twenty-first century will thus continue to exhibit the characteristic horizontal and decentralised features of international law-making identified above.

NON-CARBON ALTERNATIVES: NEW ENERGY SOURCES AND TECHNOLOGIES

Renewables

The 1990s retreat from nuclear sources of energy, combined with the need to reduce greenhouse gases from carbon-based energy sources, has led to a closer embrace of renewable sources of energy. International regulation of these sources is still in its infancy, owing partly to its traditionally relatively insignificant contribution to meeting energy demand in developed countries and partly to the absence of significant transboundary effects. Recently

[54] There has been some discussion of the creation of a 'world environmental organization' with a compulsory dispute settlement body, though politically such proposals likewise do not enjoy the widespread support of States: for proposals, see, eg G Palmer, 'New Ways to Make International Environmental Law' (1992) 86 *American Journal of International Law* 259 and G Sampson, 'The World Environment Organization and Global Environmental Governance' in W Bradnee Chambers and J Green (eds), *Reforming International Environmental Governance: From Institutional Limits to Innovative Reforms* (Tokyo, UNU Press, 2005) 124.

this picture has been changing, particularly with increased use of maritime zones for the generation of renewable energy in the hopes both of avoiding the 'Nimby' (not in my backyard) syndrome so evident in onshore developments, and to harness the immense tidal and wind energy found there. Ronne and McHarg, for example, note increasing recourse to offshore wind installations in Denmark and in the UK, amongst others.[55]

While onshore development can experience a myriad of legal obstacles arising from domestic land-use planning and environmental controls,[56] offshore developments are neither immune from public opposition[57] nor from national and international regulation. As for other maritime energy activities, the general jurisdictional framework is provided by the 1982 United Nations Convention on the of the Sea (LOSC), which clearly recognises that coastal states enjoy sovereign rights over the natural resources of their exclusive economic zone (EEZ), and 'over other activities for the economic exploitation and exploration of the zone, such as the production of energy from the water, currents and winds' (Art 56). Additional international requirements arise under existing treaty instruments including procedural requirements for a prior environmental impact assessment by States party to the 1991 Espoo Convention on the Transboundary Impact of Activities[58] and requirements for the safety of navigation[59] and of wildlife, especially for birds.[60]

[55] A Ronne and A McHarg, 'Reducing Carbon Emissions in Electricity Generation: Is the Answer Blowing in the Wind?' in Zillman *et al*, *Carbon Economy*, above n 2.

[56] Onshore wind turbines have also been unsuccessfully challenged as a nuisance impairing, inter alia, peaceful enjoyment of property rights protected under Art 1 of the First Protocol (1952) to the 1950 European Convention on Human Rights: *Fagerskiold v Sweden*, decision of 26 February 2008, available at cmiskp.echr.coe.int.

[57] O A Woolley, 'Trouble on the Horizon? Addressing Place-Based Values in Planning for Offshore Wind Energy' (2010) 22 *Journal of Environmental Law* 223.

[58] While Annex I does not list renewable energy activities beyond large hydroelectric projects, prior EIA will nonetheless be required where the project is likely to cause significant transboundary environmental impact. For example, Sweden received an Espoo notification from Germany regarding changes to an offshore wind farm license (Wikinger, formerly Ventotec Ost 2) in Germany's EEZ and participated in the eia in 2010: see www.naturvardsverket.se. The environmental impact of non-wind renewable energy systems has also recently been reviewed in the North-East Atlantic region, in particular of wave and tidal devices: see OSPAR Commission, *An Overview of the Environmental Impact of Non-Wind Renewable Energy Systems in the Marine Environment* (2006).

[59] For example in the IMO's guidance on routing systems, eg para 3.10 of IMO Recommendation A.572(14) available at www.imo.org. For general discussion of the navigational and environmental impacts of such development in the UK context, see K Scott, 'Tilting at Windmills: Regulating Wind Farm Development Within the Renewable Energy Zone' (2006) 18 *Journal of Environmental Law* 89.

[60] See eg the Bonn Convention on Migratory Species, Resolution 7.5, 'Wind Turbines and Migratory Species' (2002) available at www.cms.int. OSPAR has also addressed the environmental impacts of offshore wind farms: see 'Guidance on Offshore Wind-farms in relation to assessment of the environmental impacts of construction and best environmental practice for construction' available at www.ospar.org.

While various academic proposals have been made for an international treaty on renewables,[61] there are as yet no international instruments parallel to, for example, the EU renewables directive.[62] What is particularly striking about the EU approach is that, in addition to the emissions trading scheme, there is an effort sharing decision whereby binding (but differentiated) national targets are set for each EU member State to achieve in terms of final energy consumption from renewable sources.[63] In contrast, internationally there has been a failure even to agree aspirational targets in a soft law instrument, the non-binding Johannesburg Plan of Implementation adopted at the World Summit on Sustainable Development in 2002. While there is clearly no appetite internationally for fettering by soft or hard instrument national energy choice in terms of the mix between renewables and other forms of energy, in 2009 there was agreement to establish the International Renewable Energy Agency (IRENA) in Abu Dhabi.[64] However, its role is confined to the promotion, through information exchange and the sharing of best practices, of widespread and increased adoption and sustainable use of all forms of renewable energy.

Energy Efficiency

Despite recognition of the importance of energy efficiency in reducing energy consumption and in combating climate change,[65] there has been very little attention paid to this issue at the international level. In part this is because such measures tend to be taken at the national and local levels and rarely have transboundary consequences. That said, in 1994 the Energy Charter Treaty (ECT) was concluded with an accompanying Protocol on Energy Efficiency and Related Matters, which remains the only treaty instrument to date to address energy efficiency.[66] The preamble acknowledges the energy consumption and environmental advantages of improved energy efficiency, as well as its contribution to energy security and in improving

[61] A Bradbrook, 'The Development of Renewable Energy Through Public International Law' in Zillman *et al*, *Carbon Economy*, above n 2, and Lyster and Bradbook, *Energy Law*, above, n 33.

[62] Directive (EC) 2009/28 on the promotion of the use of energy from renewable sources, amending and subsequently repealing Directives 2001/77/EC and 2003/30/EC [2009] OJ L 140/16, available at europa.eu.int.

[63] Annex I of the Directive contains national targets for each Member State's share of energy from renewable sources, collectively to secure a share of 20% of final energy consumption from renewable sources by 2020.

[64] As of mid-July 2011, 77 States and the EU have ratified the Statute of the Agency. The first assembly of the Agency took place in April 2011 in Abu Dhabi. For further details, including the 2010 and 2011 programmes of work of the Agency, see www.irena.org.

[65] The IEA's 2009 World Energy Outlook calls for increased investment in energy efficiency and low-carbon power generation (US $10.5 trillion in the period 2010–30), above n 4; see also Barry Barton, 'The Law of Energy Efficiency' in Zillman *et al*, *Carbon Economy*, above n 2.

[66] See further Thomas Waelde *et al*, 'The Energy Charter Treaty' in Roggenkamp *et al*, *Energy Law in Europe*, above n 12.

living standards. However, its provisions are couched in largely hortatory terms, focusing on the threefold objectives of promoting energy efficiency policies, creating framework conditions to induce energy efficiency by producers and consumers through efficient and transparent energy markets, and fostering co-operation in energy efficiency (Art 1). Nor, with 51 State parties, does it have global reach. In consequence, global consideration has thus far been entirely confined to the policy level, such as the inclusion of energy efficiency among the topics addressed in 'A Framework for Action on Energy' produced in advance of the 2002 Johannesburg Summit on Sustainable Development, and in the Plan of Implementation resulting from it.

Alternative Sources of Energy: Biofuels

Enhancing energy security and diversity on the one hand, and mitigating climate change on the other, are the twin drivers behind the increased production of liquid biofuels,[67] which have been recognised as having a role to play in promoting new and renewable sources of energy for a sustainable energy future.[68] The main biomasses used in such production are sugar cane and grain corn, with second generation experimental feedstocks including rapeseed and palm oil for biodiesel.[69] This is of particular importance for the transport sector and the contribution it can make to reducing GHG emissions. However, such activities are not carbon neutral (eg the energy consumed by, and emissions from, intensive farming practices) and recently concerns have arisen regarding the environmental and developmental impact of first generation biofuels. This has been exacerbated by the requirement in the first (2001) EU renewables directive that at least 10 per cent of transport fuel must be renewable (biofuels, hydrogen, green electricity, etc). However, under the more recent 2009 renewables directive, biofuels are subject to meeting agreed sustainability criteria, including achieving minimum levels of GHG savings and that they should not be obtained from land with high biodiversity value.[70]

[67] The European Commission's working paper also identifies the need for diversification of agriculture and rural economies, mitigation of the effects of sugar reforms on developing countries, and enhancing competitiveness, as additional drivers of EU biofuel production: see European Commission, 'An EU Strategy for Biofuels: an Impact Assessment' (2006) 34 *COM* final.

[68] Biofuels were considered in the context of energy for sustainable development at the fourteenth (2006) and fifteenth (2007) sessions of the Commission on Sustainable Development available at www.un.org/esa/dsd/csd/csd_index.shtml.

[69] See Note by the Executive Secretary, *New and Emerging Issues Relating to the Conservation and Sustainable Use of Biodiversity—Biodiversity and liquid biofuel production*, UNEP/CBD/SBSTTA/12/9, 25 April 2007 and ibid, *The Potential Impacts of Biofuels on Biodiversity* UNEP/CBD/COP/9/26, 24 April 2008.

[70] Directive 2009/28, above n 62, Arts 17 (2) and (3). For thorough analysis of the Directive and the trade implications, see S Switzer and JA McMahon, 'EU Biofuels Policy—Raising

intergenerational liability, has been confronted in fashioning abandonment regimes for offshore installations, the decommissioning of nuclear plants, and with respect to offshore dumping of waste. The adaptation of existing marine environment treaties to address sub-seabed storage of CO_2, considered below, further makes the point. At the most recent COP16/CMP 6 in Cancun in 2011, the Parties decided that CCS in geological formations is eligible as project activities under the CDM *provided that* concerns expressed regarding matters such as monitoring, reporting and verification, environmental impacts, project boundaries, 'international law', 'liability' and compensation for damage caused due to seepage or leakage, are satisfactorily addressed.[82] The elaboration of modalities and procedures continues to be addressed between COP/CMPs by the SBSTA, in the light of submissions by the Parties and of a CCS Technical and Legal Experts Meeting expected to occur in Panama in autumn 2011, prior to the 35th meeting of SBSTA.[83] Significant issues remain on the table and it is doubtful whether there will be satisfactory resolution of them by the next COP/CMP in Durban in 2011. Doubtless part of the difficulty is the stage of technological development, uncertainties regarding long-term impacts and liabilities, resistance to the potential deployment of a costly technology which perpetuates carbon dependency, and issues regarding governance of the CDM.[84] It is also clear that the Technology and Financial Mechanisms agreed in principle at Cancun and to be further elaborated upon at Durban will be key to the successful deployment of CCS in developing countries in particular.[85]

International Legal Regulation of CCS

As noted above, one of the uncertainties initially identified by the CDM Executive Board was the legal status of offshore geological storage of CO_2. Both the 1996 Protocol to the London Convention and Annexes II and III to the OSPAR Convention have subsequently been amended expressly to bring defined CCS activities within their remit. This has added much needed legal clarity since, prior to these amendments, there was debate regarding the application of the 1972 London Convention and of OSPAR

[82] Decision 7/CMP.6, Carbon dioxide capture and storage in geological formations as clean development mechanism project activities, para 1, referring to the issues identified in decision 2/CMP.5, para 29.

[83] Decision 7/CMP.6, paras 2–3 (request to SBSTA and issues to be addressed); and para 5 (experts' meeting).

[84] See Decision 2/CMP.5, paras 6–15 of which are addressed to Executive Board governance issues.

[85] For fuller discussion of these mechanisms see C Redgwell, 'Facilitation of Compliance' in J Brunnee, M Doelle and L Rajamani (eds), *A Compliance System for the Post-2012 Climate Change Regime* (Cambridge, Cambridge University Press, 2011).

to offshore CCS, with application turning on questions such as whether the matter stored is 'waste' for the purposes of the convention, and whether the convention addresses geological disposal in the seabed.[86]

Owing to vigorous campaigning by Australia in particular, the first meeting of the contracting Parties following the entry into force of the Protocol in 2006 adopted an amendment the legal effect of which is to permit carbon capture and storage—specifically sub-seabed disposal of CO_2—with effect from 10 February 2007.[87] The amendment is described as 'part of a suite of measures to tackle climate change and ocean acidification including, first and foremost, the need to further develop low carbon forms of energy'.[88] Consistent with the 'prohibited unless permitted approach' of the Protocol (Art 4), the amendment adds to the list of exceptions (ie permitted activities) in Annex I 'carbon dioxide streams from carbon dioxide capture processes for sequestration'. This is supplemented by further guidance on the means for sub-seabed carbon sequestration.[89] Under this amendment, 'carbon dioxide streams... may only be considered for dumping, if: (1) disposal is into a sub-seabed geological formation; and (2) they consist overwhelmingly of carbon dioxide. They may contain incidental associated substances derived from the source material and the capture and sequestration processes used; and (3) no wastes or other matter are added for the purpose of disposing of those wastes or other matter'.[90] By implication, given the narrow scope of the permitted exception in Annex I, disposal or storage of CO_2 in the water column is prohibited.

The regional OSPAR Convention has adopted a similar approach for the regulation of CCS activities in the North-East Atlantic, but has taken a step further. Like the 1996 Protocol, it does so against a convention framework adopting a 'prohibited unless permitted' approach.[91]

[86] For in-depth treatment of these issues under the London Convention and OSPAR prior to amendment, see R Purdy and R Macrory, *Geological Carbon Sequestration: Critical Legal Issues* (Tyndall Centre for Climate Change Research Working Paper, 2004) available at www.tyndall.ac.uk/content/geological-carbon-sequestration-critical-legal-issues.

[87] The 1996 Protocol contains expedited procedures for amendment of its annexes, which become binding on all parties within 100 days of adoption unless a party expressly opts out of the amendment. This is called the 'tacit amendment' procedure which is an innovation of IMO conventions in particular. No party has exercised the right to opt-out of this amendment. However, for States still party to the 1972 Convention, its provisions will continue to bind, with the attendant legal uncertainties regarding its application to CCS—a further incentive to ratify the 1996 Protocol.

[88] See London Convention homepage at www.imo.org.

[89] 'Specific Guidelines for Assessment of Carbon Dioxide Streams for Disposal Into Sub-seabed Geological Formations' were adopted by the 29th meeting of the Parties in November 2007.

[90] Para 4, Annex I, 1996 Protocol.

[91] It entered into force on 25 March 1998. There are presently 16 parties to the OSPAR Convention, including the EC. For text see www.ospar.org.

OSPAR Decision 2007/1 expressly prohibits the placement (encompassing disposal) of CO_2 streams in the water column or on the seabed because it 'is likely to result in harm to living resources and marine ecosystems and is thus neither a viable solution with regard to mitigating climate change nor compatible with the aims of the Convention'.[92] It is also stressed that CCS is but one in a portfolio of options for reducing CO_2 emissions—other options including energy conservation (demand restraint), renewables and improved energy efficiency.[93] A second decision, Decision 2007/2 on the Storage of Carbon Dioxide Streams in Geological Formations,[94] provides the legal framework for the storage of CO_2 streams in sub-soil geological formations, to be read in conjunction with 'Guidelines for Risk Assessment and Management of Storage of CO_2 Streams in Geological Formations' which are subject to regular review as scientific knowledge of the environmental risks of CCS develops. The purpose of the Decision is to ensure permanent containment with no significant adverse consequences for the marine environment, human health and other legitimate uses of the maritime area (para 2.1). Storage must be subject to permitting or approvals processes by the competent authorities which must at a minimum contain the following:

1. a description of the operation, including injection rates;
2. the planned types, amounts and sources of the CO_2 streams, including incidental associated substances, to be stored in the geological formation;
3. the location of the injection facility;
4. characteristics of the geological formations
5. the methods of transport of the CO_2 stream;
6. a risk management plan that includes:
   i. monitoring and reporting requirements;
   ii. mitigation and remediation options including the pre-closure phases; and
   iii. a requirement for a site closure plan, including a description of post-closure monitoring and mitigation and remediation options; monitoring shall continue until there is confirmation that the probability of any future adverse environmental effects has been reduced to an insignificant level.[95]

[92] OSPAR Decision 2007/1 to Prohibit the Storage of Carbon Dioxide Streams in the Water Column or on the Sea-Bed, OSPAR 07/24/1-E, Annex 5, 25–29 June 2007, preamble, 7th indent. The environmental impact of subsea disposal is considered in a 2006 OSPAR report, *Placement of CO_2 in Subsea Geological Structure While the General Impact of Increased CO_2 on the Marine Environment Is Addressed in Effects on the Marine Environment of Ocean Acidification Resulting From Elevated Levels of CO_2 in the Atmosphere* (2006).

[93] Ibid, 4th Indent, and OSPAR Press Statement, 'New Initiatives on CO_2 Capture and Storage and Marine Litter' (28 June 2007) 2.

[94] OSPAR 07/24/1-E, Annex 6.

[95] Ibid, para 3.2 (emphasis in original). These conditions address the key concerns identified in, inter alia, the CDM Executive Board's Recommendation discussed above.

In consequence of these amendments, it is clear that sub-seabed geologic storage of CCS is regulated both globally (London Convention) and at the regional level (OSPAR). Other regional developments may be expected, for example under United Nations Environment Programme's regional seas programme, following in the footsteps of OSPAR. The impetus for these developments is twofold: (i) climate change mitigation; and (ii) protection of the marine environment from acidification arising from carbon-based activities in general, and from the risks of CCS in particular. Overarching these pollution and/or region specific arrangements is, of course, the general 1982 LOSC. While it does not include explicit permissions or prohibitions on the disposal of specific substances in the oceans, the detail of which is left to these more specialised global and regional instruments,[96] the general obligations to protect and preserve the marine environment are clearly applicable.

CONCLUSIONS

This chapter has examined international legal responses to new and emerging technologies which are increasingly resorted to as part of a suite of climate change measures. In each case other benefits arise, such as enhancing energy security, as well as creating risks, particularly to the environment. Offshore renewables present the strongest case for 'win-win' measures in the context of extensive existing regulation of the marine environment which has been able flexibly to adapt to address, and thus to mitigate, the negative impacts of such development.[97] This flexibility is also evident in the swift response to the uncertainties created over the legal status of sub-seabed sequestration of CO_2 through treaty adaptation and amendment. Biofuels are a further example of the integration of new and emerging issues into existing treaty frameworks, with the integration of biofuels into the agricultural and forestry biodiversity work streams of the CBD. The sustainability concerns expressed, inter alia, by the CBD clearly influenced the second generation regulation of renewables in the EU with the 2009 Directive embedding explicit sustainability criteria. In the climate change context, the process of integrating CCS within the CDM has been a slower and more complex process. There are a variety of reasons for this which relate not so much to the suitability or adaptability of existing legal rules

[96] For the relationship between the LOSC and these other dumping treaties, see C Redgwell, 'From Permission to Prohibition: The 1982 Convention on the Law of the Sea and Protection of the Marine Environment' in D Freestone, R Barnes and D Ong (eds), *The Law of the Sea: Progress and Prospects* (Oxford, Oxford University Press, 2006).

[97] Indeed, this is a hallmark of many environmental treaties which are required to evolve over time as scientific knowledge changes: see Redgwell ibid.

but rather to the inherent complexity of climate change issues,[98] the cost of the technology, its environmental impact and—as for geoengineering methods—the concern that the technology does not address the underlying 'carbon addiction' and permits business as usual.

What is striking in each of these cases is the clear reluctance of States to place responses within a wider framework of commitment to a particular energy mix and to constrain national energy choice. Each of the developments above is an example of isolated 'problem solving' under existing international instruments, often buttressed by prior developments in national and regional (EU) law—both biofuels and CCS make the point. To adapt Sand's observation of developments in the climate change context, what international energy law lacks in the twenty-first century is not the necessary legal tools but in fact a 'grand narrative'.[99]

[98] It has been suggested that recent climate change developments signal the failure of multilateralism as one model of global governance: D Bodansky and E Diringer, *The Evolution of Multilateral Regimes: Implications for Climate Change* (Arlington VA, Pew Centre on Global Climate Change, 2010).

[99] P H Sand, 'The Evolution of International Environmental Law' in D Bodansky, J Brunnee and E Hey (eds), *The Oxford Handbook of International Environmental Law* (Oxford, Oxford University Press, 2007) 40.

3

UK Climate Change Litigation: Between Hard and Soft Framing

CHRIS HILSON

INTRODUCTION

CLIMATE CHANGE LITIGATION has received considerable legal academic attention in the US (particularly in the wake of Hurricane Katrina)[1] and also within international law (for example in relation to human rights petitions by the Inuit,[2] and actions by small island states[3]).[4] While EU climate change litigation is now the subject of a growing literature,[5] there has been much less academic research on UK examples of

[1] Which has given rise to climate change litigation in the US in, eg *Comer v Murphy Oil*, US Court of Appeals 5th Circuit, No 07-60756, 22 October 2009 and (dismissing the appeal) 28 May 2010, in which residents along the Mississippi coast were seeking damages based on the torts of public and private nuisance, trespass and negligence, against a number of industries, alleging that they had contributed to climate change and thus to the weather events involved with Hurricane Katrina.

[2] Petition to the Inter American Commission on Human Rights Seeking Relief from Violations Resulting from Global Warming Caused by Acts and Omissions of the United States, 7 December 2005.

[3] See, eg the request filed by Micronesia with the Czech Government for a transboundary EIA of a major Czech coal power station. The Czech Republic is a party to the Espoo Convention on Environmental Impact in a Transboundary Context, although Micronesia is not, which would prevent the latter from judicially enforcing the Convention. See further solveclimate.com/blog/20091218/island-nation-citing-climate-change-threat-goes-after-czech-coal-plant.

[4] From a rapidly developing literature, see further, eg H Osofsky, 'Climate Change Litigation as Pluralist Legal Dialogue' (2007) 26 *Stanford Environmental Law Journal* 181 and 43 *Stanford Journal of International Law* 181 (joint issue); Special Issue on Climate Change Litigation (2009) 18 (2) *Review of European Community and International Environmental Law* 119–71; W Burns and H Osofsky, *Adjudicating Climate Change: State, National and International Approaches* (Cambridge, CUP, 2009).

[5] For recent work on European climate change litigation (largely centred on the EU emissions trading scheme), see, eg N Singh Ghaleigh, 'Emissions Trading Before the European Court of Justice: Market Making in Luxembourg' in D Freestone and C Streck (eds), *Legal Aspects of Carbon Trading: Kyoto, Copenhagen and Beyond* (Oxford, OUP, 2009); A-S Tabau and S Maljean-Dubois, 'Non-Compliance Mechanisms: Interaction Between the Kyoto Protocol System and the European Union' (2010) 21 *European Journal of International*

such litigation. The aim of this chapter is to use developing UK litigation associated, in particular, with coal fired power stations and airport expansion, as case studies for examining the way in which the climate change movement and the state authorities and media have engaged in a subtle 'framing' war.

The chapter's core argument is that framing has been central to many of the movement actions and state and media reactions. Thus, the police, prosecution authorities and (mostly right-wing) media reporting of climate change protest have tended to apply a 'hard' frame, whereby the protesters are constructed as at best criminals and, at worst, extremists or terrorists who are threatening not only law and order but also, for example, industrial and aviation security and personal freedoms. The climate change movement has then attempted to reframe using 'soft' counter-frames seeking to construct itself as, inter alia, reasonable, citizenly, acting in the public interest, and upholding the rule of law.

FRAMING

Framing, which is a core concept within the literature on social movements, involves a purposeful struggle over ideas and meanings,[6] played out between movements, institutions of power including the media, and counter-movements. It is a *contentious* process in so far as 'it involves the generation of interpretive frames that not only differ from existing ones but that may also challenge them.'[7] There are a number of respects in which framing can be said to be relevant to climate change litigation. First, legal cases which would previously have been 'just' about, for example, open-cast coal mining's effect on the natural landscape,[8] or airport noise,[9] are now also actively constructed as claims involving climate change. This is really a point about the legal framing *of* cases: many environmental cases are increasingly framed using the language of climate change. The reasons for this are many: given the currency of climate change as one of the key political issues of our time, it may be thought to add weight to the overall

Law 749; N Ghaleigh, '"Six-Honest Serving Men': Climate Change Litigation as Legal Mobilization and the Utility of Typologies' (2010) 1 *Climate Law* 31; and J van Zeben, 'The European Emissions Trading Scheme Case Law' (2009) 18 *Review of European Community and International Environmental Law* 119.

[6] R Benford and D Snow, 'Framing Processes and Social Movements: An Overview and Assessment' (2000) 26 *Annual Review of Sociology* 611, 613.

[7] Ibid 614.

[8] See, eg the Australian 'Anvil Hill' case, *Gray v Minister for Planning* [2006] NSWLEC 720—a case challenging an EIA decision on the open-cast Anvil Hill coal mine, which had failed to consider the indirect greenhouse gas emissions that would result from the extracted coal being burnt.

[9] Eg Barbone, n 20 below.

case; in airport and also incineration cases,[10] it can be seen as a means of adding a global frame to support what might otherwise be characterised by critics as a local, Nimbyish protest; however, in other cases such as those involving challenges to wind farms, climate change frames may well be used as a way of attempting to undermine Nimby-based, local impact arguments.[11]

However besides frames used by the movement *in* litigation proceedings, one must also consider those employed by the movement in non-court settings (for example in protests or direct action). The present chapter examines not so much frames adopted within particular strategies such as litigation, but rather strategies themselves as a form of framing activity. In this respect, the chapter shares much in common with McAdam's study of the US civil rights movement, in which he argues that studies of framing have tended to pay too much attention to ideational aspects of framing and insufficient attention to examining tactics as frames.[12] As McAdam states, in some senses movement actions do speak, or signify, louder than words.[13]

Applying this insight to the current chapter, to adopt direct action as a tactic or strategy is to engage in a particular form of public theatre which invites the audience—including the police, prosecution authorities, judges, juries, the media and through them the public—to ascribe a particular meaning to it. The police and right-wing media response to environmental direct action is often to construct or frame such behaviour as at best 'criminal' (involving, for example criminal damage, aggravated trespass and so on) and at worst an example of (eco-) terrorism. These are what can be characterised as 'hard' frames, with criminality at one end of the hardness spectrum and extremism and terrorism at the far end. Policing tactics can themselves be seen as attempting to frame or construct the objects of that policing in a certain way. Therefore, using, say, kettling techniques to herd and contain protesters together, or police powers of stop-and-search under

[10] Eg the Newhaven incinerator case (n 10) and the *Barbone* Stansted airport case (n 39 below).

[11] *Cf* Awel Aman, '*Tawe Cyfngedig v The National Assembly for Wales*' [2007] 69 (Oct) *All England Direct Law Reports (Digests)*, which involved a (charity) developer challenge to a decision refusing planning permission for a wind farm in a notable landscape area. Although, climate change framing is not fully explicit in the case, climate change does feature indirectly: the decision was (unsuccessfully) challenged on the basis that the planning inspector had mistakenly applied the relevant policy (ie that he had found that large output wind farms could not be built other than in pre-designated areas, which the current site fell outside, and had thus failed to take into account the possibility of environmental, including climate change, reasons which could exceptionally justify this).

[12] D McAdam, 'The Framing Function of Movement Tactics: Strategic Dramaturgy in the American Civil Rights Movement' in D McAdam, JD McCarthy and MN Zald (eds), *Comparative Perspectives on Social Movements: Political Opportunities, Mobilizing Structures and Cultural Framings* (Cambridge, CUP, 1996) 339–40, 354.

[13] Ibid, 341.

legislation designed to deal with terrorism, is to try to apply a hard frame to protesters. Where they occur, arrest and prosecution as tactics similarly seek to inscribe a particular hard, criminal frame on climate change protesters.

Unlike the UK anti-roads movement of the 1990s, which became largely associated with the hard frames of direct action, the UK climate change movement has employed a range of softening counter frames in what can be seen as an attempt by the movement to seem lawful, responsible, citizenly, reasonable and acting to protect the public interest. How then has this attempt at soft counter-framing operated?

Climate Camps

In essence, the climate change movement has employed a range of strategies—other than direct action—which are clearly the opposite of criminality and disorder. The use of Climate Camps, orchestrated by the eponymous social movement organisation, provides a good example of the employment of soft strategies and thus soft frames. Such Camps—which started in 2006 at the Drax coal fired power station, and were subsequently held in 2007 at Heathrow, in 2008 at the Kingsnorth power station, and in 2009 in both the City (at the G20 Summit) and at Blackheath in London—are centred not only on direct action, which tends to be played down by the Camps, but also education, sustainable living[14] and movement building.[15] The hard frame of direct action is thus enveloped by a range of clearly reasonable and constructive softer educational and lifestyle frames. In other words, the Camps themselves are framed not as examples of confrontational protest, but rather ones that the authorities could not possibly feel threatened by or seek an excuse to close down, without themselves appearing destructive, mean-spirited and unreasonable.

Airplot!

Another example of soft framing can be found with the purchase of a half-football pitch sized parcel of land near Heathrow by a coalition of climate change protesters consisting of Greenpeace, the actor Emma Thompson, the impressionist Alistair McGowan and the (then) prospective Conservative Parliamentary Candidate and former Editor of *The Ecologist*, Zac Goldsmith.[16] Greenpeace, in a campaign known as 'Airplot!',

[14] With, for example, hay bale toilets. See further T Omond, *Rush! The Making of a Climate Activist* (London and New York, Marion Boyars, 2009) 44.

[15] See www.climatecamp.org.uk/about.

[16] news.bbc.co.uk/1/hi/england/london/7825169.stm.

subsequently organised the creation and giving away of further beneficial interests in the land, near the village of Sipson, which would be bulldozed to make way for the planned third runway.[17] These interests were initially handed to politicians including local Labour MP John McDonnell, Liberal Democrat MP Susan Kramer and Conservative front bencher Justine Greening as well as to environmentalist George Monbiot and climate scientist Dr Simon Lewis. However, Greenpeace then invited anyone to sign up for such an interest via its website. This spreading of interests in the land could be characterised as an example of 'anticipatory' climate change litigation in that, in the event of a third runway getting the go-ahead, it clearly anticipates a reactive litigation battle via the law of compulsory purchase. The tactic of creating multiple interests can also be seen as a form of legal 'monkeywrenching'. Monkeywrenching—involving actions ranging from tree-spiking to bulldozer sabotage—was a tactic employed by the anti-roads movement to obstruct and delay road building.[18] In the Heathrow-Sipson context, multiplying interests in land becomes a similar, more intangible, legal form of obstruction and delay, albeit a more reasonable one because the behaviour is clearly lawful (and indeed artfully so). As such, it clearly represents a 'soft' strategy which is again poles apart from the hard framing of direct action.

Proactive and Reactive Litigation

Litigation is of course a further type of strategy. It is worth drawing a distinction at this point between proactive and reactive litigation. Proactive litigation, in which the climate change movement is the one bringing proceedings, has featured as a strategy on a number of occasions. Many challenges have related to airport expansion. There have been judicial review claims or statutory appeals brought in relation to City Airport,[19] Stansted[20]

[17] www.greenpeace.org.uk/blog/climate/join-plot-stop-airport-expansion-20090112.

[18] On monkeywrenching, see D Foreman and B Haywood, *Ecodefence: A Field, Guide to Monkeywrenching* (Chico, CA, Abbzug Press, 1993).

[19] *R (Griffin) v Newham London Borough Council* [2011] EWHC 53 (Admin). The claim (supported by Friends of the Earth legal expertise within its Rights & Justice Centre) was brought by Anne-Marie Griffin, Chair of 'Fight the Flights', a local campaign group concerned about the effect of London City Airport on local residents and the climate. The claimant's challenge was based, inter alia, on the failure by Newham, the local planning authority responsible for giving the go-ahead to expansion of City Airport, to take account of a change in Government policy on climate change and aviation. The High Court dismissed the claim, with Pill LJ ruling that there had been no relevant change in Government policy which the council ought to have considered.

[20] *Barbone v Secretary of State for Transport* [2009] EWHC 463 (Admin) [2009] Env LR D12. This was a statutory appeal brought by the campaign group Stop Stansted Expansion (SSE) under s 288 of the Town and Country Planning Act 1990. The application against the decision to grant planning permission for the expansion of Stansted Airport was based, inter

and Heathrow.[21] These have involved the climate change movement in varying degrees, ranging from acting as joint claimants in the Heathrow *Hillingdon* case, through acting as legal representative in the City Airport case, to the Stansted statutory appeal[22] involving the campaign group Stop Stansted Expansion (SSE), which has significant informal ties with the climate change movement in the shape of some of the key environmental NGOs such as Greenpeace and Friends of the Earth (FoE). Beyond airport cases, there are judicial review challenges that have been brought by FoE in relation to fuel poverty[23] and incineration,[24] and there has also been an action brought by the World Development Movement, PLATFORM and People & Planet against the UK Treasury for allowing part-nationalised

alia, on the failure to take proper account of its effect on greenhouse gas emissions as a material consideration. The application, on this and other grounds, was dismissed. *Cf* the earlier judicial review claim brought by the Chairman of SSE, a member of HACAN, and a number of local authorities, *R (London Borough of Wandsworth) v Secretary of State for Transport* [2005] EWHC 20 (Admin) 2005 WL 353363, which involved a challenge to the Government's 2003 Aviation White Paper and did not include explicit arguments relating to climate change. The challenge was in part successful: although the decision-making leading to the White Paper was declared generally lawful, aspects of the White Paper's policy on Stansted and Luton expansion were held to be unlawful.

[21] *R (London Borough of Hillingdon) v Secretary of State for Transport* [2010] EWHC 626 (Admin)—a multi-claimant judicial review challenge to the third runway at Heathrow brought by a number of local authorities and also a range of pro-climate environmental NGOs including Greenpeace and WWF. The case was arguably a success for both sides, with Carnwath LJ finding against the Government for its reliance on the 2003 White Paper when its climate policy had clearly moved on since then, but at the same time being unwilling to quash the decision to confirm policy support for the third runway on the basis that the procedure under the new 2008 Planning Act would inevitably address the relevant issues.

[22] *Barbone* (n 20).

[23] *R (Friends of the Earth and Others) v Secretary of State for Energy and Climate Change* [2009] EWCA Civ 810; [2010] PTSR 635. The High Court judgment is reported as *R (Friends of the Earth, Help the Aged) v Secretary of State for Business Enterprise and Regulatory Reform, Secretary of State for Environment, Food and Rural Affairs* [2008] EWHC 2518 (Admin); [2009] Env LR D6. Although climate change framing does not appear on the face of the legal judgment itself, it is apparent in the associated press release, with Government failures on energy efficiency providing the bridge between climate change and fuel poverty: www.foe.co.uk/resource/press_releases/fuelpoverty_courtcase_06102008.html.

[24] *R (Lewes District Friends of the Earth) v East Sussex* CC [2008] EWHC 1981 (Admin); [2009] Env LR 11. Like the FoE case in n 23 above, this was a case where climate change appeared in a press release (drawing explicit attention to a preference for recycling over incineration, in part on climate change grounds) rather than on the face of the judgment. For the press release, see www.foe.co.uk/resource/press_releases/newhaven_incinerator_15072008.html. There was a separate, earlier judicial review challenge to the Newhaven pollution permit, which did directly raise climate change issues (the Environment Agency's failure properly to assess CO_2 emissions from the plant), but which was settled by the Agency (meaning there is no judgment): *R (Day) v Environment Agency* (2007) and the press report of the case, 'Incinerator Pollution Permit Withdrawn', Sussex Express, 6 December 2007. Although incineration is not the focus of the current chapter, for a similar, settled case, see also *R (HOTI) v Environment Agency* (2008) which involved a climate change challenge to an incinerator in Hull (for details, see www.39essex.co.uk/resources/cases.php?id=1382). I am grateful to Phil Michaels for bringing these cases to my attention.

Royal Bank of Scotland to invest in climate-unfriendly projects in breach of Government policy requiring climate change impacts to be considered.[25]

This proactive litigation as a strategy is again an attempt at soft framing in so far as judicial review involves the ideal of upholding legality and the rule of law and in that sense is a thoroughly reasonable enterprise. However, proactive litigation has not received anything like as much media attention as reactive litigation, which arises from and is thus intimately linked with direct action as a strategy. With reactive litigation, what tends to happen is that protesters engage in what the police are likely to regard as unlawful criminal activity (such as criminal damage). Thus, for example, in October 2007 at the Kingsnorth coal-fired power station in Kent, activists who would become known as the 'Kingsnorth 6', scaled the power station chimney and painted 'Gordon' on it in a message to the UK Prime Minister Gordon Brown about climate change.[26] The six were charged with causing £30 000 of criminal damage to the chimney but, in September 2008, were acquitted by a majority verdict of the jury at Maidstone Crown Court after raising a defence under section 5 of the Criminal Damage Act 1971 of lawful excuse.[27] This defence to a charge of criminal damage—which would normally apply to, for example, a neighbour breaking into a house to put out a fire—was applied here to the effect that the damage to the chimney was needed to prevent greater damage being caused by climate change to property all around the world.

This is an example of reactive litigation in that the protesters had not themselves brought proceedings, as in proactive judicial review, but rather used the criminal proceedings brought against them as another avenue for getting their protest message heard.[28] Judicial review was characterised above as an example of soft framing. With reactive litigation, the initial direct action—being seemingly unlawful, obstructive and so on—engages a hard frame by the authorities, but then an attempt at soft counter-framing is made via the rational climate change justification put forward as a defence during the subsequent (criminal) court proceedings. In the Kingsnorth case, with a

[25] peopleandplanet.org/ditchdirtydevelopment/legalchallenge. However, permission to bring judicial review proceedings in this case was refused by Sales J in October 2009. *R (People & Planet) v HM Treasury* [2009] EWHC 3020 (Admin); http://peopleandplanet.org/ditchdirtydevelopment/legalchallenge/legaldocuments.

[26] The intended message was 'Gordon Bin It', but the protesters were interrupted before being able to complete this.

[27] Like most routine criminal cases heard in Crown and magistrates' courts, this case is not reported in the law reports.

[28] On reactive litigation, see C Harlow and R Rawlings, *Pressure Through Law* (London, Routledge, 1992) 162; L Vanhala, 'Anti-Discrimination Policy Actors and Their Use of Litigation Strategies: The Influence of Identity Politics' (2009) 16 *Journal of European Public Policy* 738. Reactive litigation has been a key strategy within the UK anti-nuclear movement—see further C Hilson, 'Framing the Local and the Global in the Anti-Nuclear Movement: Law and the Politics of Place' (2009) 36 *Journal of Law and Society* 94.

jury acquittal based on a defence that their action was, in effect, lawful, the reasonableness of their action was inscribed by law and—for the jury at least and arguably wider, with the jury representing a type of public barometer—the soft replaced the hard. Far from appearing as lawbreakers, the protesters become potentially heroic figures who were literally saving the world.

Examples of soft framing like this have to be set against those of hard (often counter-) framing from the authorities. It has, for example, been reported that, after the first Climate Camp at Drax in 2006, which involved some incidents of direct action, a North Yorkshire police report referred to the incident as 'the first time domestic extremism took place against national infrastructure in the county'.[29] This effectively involved the police framing direct action not just with a hard, criminal frame, but with the even harder frame of domestic extremism.

The reference to national infrastructure is important here. In an example of protest against airport expansion, climate activists cut through the perimeter fence at Stansted airport and created a fort out of security fencing near to the runway. As a result, the airport was closed for five hours and there were knock-on delays for a further three days. Both airports and power stations are key elements of the economic infrastructure and, particularly post-9/11, have been sites of anxiety about terrorist attack. The fact that they are shared targets for terrorists and environmental protesters[30] arguably makes it easier for critics of direct action to place such action within a very hard domestic extremist or terrorist frame.[31] At the very least, it makes it easier for media reporting to focus on the security breach per se, on a 'what if' it had been terrorists basis, with little space given over to the climate change justification for the action.[32] Alternatively, as in the *Garman* Heathrow injunction case,[33] one finds statements in the judgment about police and security attention being directed on protesters and thus away from the terrorist threat:

> Police who would ordinarily be deployed on other duties at the airport and elsewhere will be required instead to deal with protestors. They will be deflected from

[29] R Evans, P Lewis and M Taylor, 'How Police Rebranded Lawful Protest as "Domestic Extremism"' www.guardian.co.uk/uk/2009/oct/25/police-surveillance-protest-domestic-extremism.

[30] For a media report making a link, in relation to the breach of security, between the climate activists and potential terrorist attack, see, eg 'Security Fears Follow Stansted Breach' www.independent.co.uk/news/uk/home-news/security-fears-follow-stansted-breach-1057706.html.

[31] On the danger of public elision of direct action with terrorism, see, eg 'Do Direct Action Protests Cause More Harm Than Good?' www.guardian.co.uk/news/blog/2008/dec/13/activists.

[32] Ibid. See also Omond, above n 14, 90.

[33] *Heathrow Airport Ltd v Garman* [2007] EWHC 1957 (QB) (noted in J Thornton, 'UK Law—Significant Environmental Cases 2007–08' (2008) 20 *Journal of Environmental Law* 293, 313–4). The case involved the operators of Heathrow seeking an injunction against Plane Stupid and other protestors who were seeking to engage in protest at Heathrow itself during the Heathrow Climate Camp (located near to Heathrow). While the injunction (which was granted) prevented significant disruption at Heathrow, it was not directed at the Climate Camp itself.

> their ordinary duties and—a matter of considerable concern—from carrying out their vital role in protecting the public (in particular the travelling public) from terrorist attack. There is a risk that a terrorist group might use the disruption caused by the protestors to perpetrate an attack on the airport with disastrous consequences.[34]

The above Stansted protest is also instructive in comparison with the Kingsnorth case discussed earlier. Like Kingsnorth, it is an example of reactive litigation in that the Stansted protesters were prosecuted for aggravated trespass and used the court proceedings to publicise the climate change motivation for their actions. For example, defence counsel Benjamin Newton stated that the group 'felt government policy was directly contrary to meeting the country's international obligations to mitigate climate change and that those policies were going to make us closer to the tipping point'.[35]

However, while this rationalisation can be seen as an attempt to soften the hardness of the direct action frame, since the case was tried in a magistrates' court[36] rather than a Crown Court with jury, the protesters were never likely to achieve the ultimate softening in court via a Kingsnorth-type acquittal.[37] Magistrates are, historically, much less favourably inclined to protesters than juries.[38] Neither was the attempt at softening helped by the fact that—unlike power station protests which have so far not managed to shut down supply and hence inconvenience the public—the Stansted protest led to the cancellation and delay of numerous flights, including many operated by the budget airline Ryanair.[39] The inconvenience of this led to considerable anger on the part of travellers whose flights were affected and the action attracted criticism in the mainstream right-wing media based on the selfishness and middle-class self-righteousness of the activists.[40] For frame softening to work, direct action must be constructed as in some sense justified by the greater good; however, the lesson of Stansted is that significant

[34] Ibid, para 112 of Swift J's judgment.

[35] R Williams and J Vidal, 'Stansted Protesters Sentenced To Community Service Amid Threat of £2m Damages Claim' *The Guardian*, London, 8 January 2009, 4.

[36] Harlow Magistrates' Court.

[37] Other airport expansion cases heard by magistrates (and similarly resulting in convictions despite climate defences) include, eg the two 'Manchester Airport on Trial' cases www.manchesterairportontrial.org/ of December 2010 and February 2011.

[38] See, eg Hilson, above n 28, 100–01; S Roseneil (ed), *Disarming Patriarchy: Feminist and Political Action at Greenham* (Open University Press, 1995) 108.

[39] Which was threatening to bring civil claims not only against BAA for failing to provide appropriate security cover at Stansted but also against individual protesters, above n 35.

[40] See, eg L Haywood, 'Upper Crusties' www.thesun.co.uk/sol/homepage/news/2018479/Posh-protesters-cause-airport-mayhem.html; D Millward, 'Plane Stupid's Antics at Stansted Airport are Plain Selfish' www.telegraph.co.uk/comment/3687100/Plane-stupids-antics-at-Stansted-airport-are-plain-selfish.html. Although such criticism was principally confined to the mainstream right-wing press, there was some critical comment in the minority, online, left-liberal media, which saw the protest as elitist and anti-'Chav' or working class: B O'Neill, 'Class Hatred at Stansted Airport' www.spiked-online.com/index.php/site/article/6008/.

public disruption makes it less likely that actors such as magistrates, the public and the media will widely adopt such a construction. In such circumstances, the hard, criminal frame, or even the harder extremist frame,[41] are more likely to stick—both in court and among the wider public.

In another incident at Drax in June 2008, climate change protesters hijacked a train carrying coal to the power station, hoisting a banner saying 'Leave it in the ground' and shovelling coal off the wagons.[42] In their subsequent trial in July 2009,[43] which was before a jury, the so-called 'Drax 22'[44] had wanted to raise a Kingsnorth-like, climate change-based defence of necessity to the charge of obstruction of the railway,[45] only to have this defence and evidence of climate change ruled out by the presiding judge, HHJ Spencer, in a pre-trial hearing:[46]

> evidence concerning the burning of fossil fuels and global warming is inadmissible. To rule otherwise would allow these defendants to hijack the trial process just as surely as they hijacked the coal train.[47]

The consequence of this was that, unlike at the earlier Kingsnorth trial, the Drax defendants were unable to call expert witnesses like Professor James Hansen,[48] in their defence. Neither were they meant to point to global warming to justify their action at all during their defence. Although the judge apparently intervened on a number of occasions to prevent the latter, in the end he in fact appears to have let a fair amount of comment on climate change through.[49]

Although it is not possible to establish a definite causal connection between the guilty verdicts and the judge's pre-trial ruling and later directions to the jury to ignore any evidence they had heard on climate change,

[41] J Lewis, 'Secret Police Unit Set Up to Spy on British 'Domestic Extremists'' www.dailymail.co.uk/news/article-1138755/Secret-police-unit-set-spy-British-domestic-extremists.html.

[42] A message which angered some in the Labour movement, given the precarious position of the miners—see, eg www.workersliberty.org/story/2009/06/30/defend-drax-29.

[43] *R v Bard*, 1 June 2009, unreported.

[44] Some reports refer to the 'Drax 29'. However, only 22 pleaded not guilty at the trial at Leeds Crown Court, with 5 having previously pleaded guilty and a further two unable to attend trial due to ill-health (M Wainwright, 'Drax Protesters Found Guilty of Obstructing Coal Train' www.guardian.co.uk/environment/2009/jul/03/drax-coal-train-trial-guilty).

[45] On the difference between the defences available in the Kingsnorth and Drax cases, see further M Schwarz, 'The Drax 29 and the Kingsnorth 6: Different Defences, Different Outcomes' ELFline (Spring/Summer 2010).

[46] After the embarrassing Kingsnorth acquittals, the Government's Attorney General was considering making a reference to the Court of Appeal to appeal the point of law involving the use of justification defences by protesters (www.greenpeace.org.uk/blog/climate/government-planning-take-kingsnorth-defence-away-juries-20081217). However, in the end, no reference was made. The judge's own initiative in the Drax case essentially achieved much the same result, if only on a one-off, non-precedent setting basis.

[47] B Stewart, 'Drax Trial Held in a Climate of Injustice' www.guardian.co.uk/commentisfree/2009/jul/03/drax-trial-climate-change-injustice.

[48] The well-known climate scientist.

[49] Wainright, above n 44.

for present purposes what is significant about the case is that the judge effectively denied the protesters the formal ability to replace the hard frame of their original direct action with the soft frame of reasonableness and lawful justification. The local chief Crown Prosecutor was similarly keen to ensure the correct 'hard' framing of the protesters' action in his post-trial statement when expressing that '[t]his was not a peaceful demonstration about the environment, but a well-planned and executed crime.'[50]

Another example of hard framing in the context of power station protest can be found with the pre-emptive, night-time arrest of 114 protesters camping out at a Steiner School near to the Ratcliffe-on-Soar power station owned by E-ON, the owners of Kingsnorth. Although bolt-cutters and locking-on type equipment was apparently found, suggesting that prima facie unlawful interference with the power station was being planned, media coverage of and commentary on the arrests highlighted the terrorist-like framing of the police raid and detention of suspects.[51] As the local Labour MP Alan Simpson stated:

> I am absolutely baffled by the sheer scale of the police operation ... It was very Orwellian. What we saw was over-the-top, smash-and-grab, SAS-style; pre-emptive policing that was massively disproportionate to what was happening on the ground. The scale of policing was what you would expect to be used for a terrorist event or the break-up of a major crime syndicate, not to stop an environmental protest.[52]

What was particularly striking about the arrests was first the scale of the operation, involving over 200 officers from five forces[53] and a fleet of police vans to take the suspects away; secondly, their timing at night, associated as this is with major criminality or terrorism; and finally, their pre-emptive nature, effectively snuffing out any form of lawful protest before it had begun by associating the whole event with an unlawful, conspiratorial quality.

The incident resulted in two subsequent jury trials. In the first trial, 20 of the protesters admitted that they intended to close down the power station, but nevertheless pleaded not guilty to conspiracy to commit aggravated trespass because they intended to raise climate change-based defences of necessity and also the use of reasonable force to prevent crime under section 3 Criminal Law Act 1967. In a pre-trial hearing,[54] Flaux J cast doubt on the approach adopted in the Drax, *Bard* case by HHJ Spencer, preferring

[50] H Carter, 'Drax Coal Train Hijackers Sentenced' www.guardian.co.uk/environment/2009/sep/04/drax-protesters-sentenced.

[51] M Taylor and P Lewis, 'Nottingham Power Station Protesters 'Treated Like Terrorists'' www.guardian.co.uk/environment/2009/apr/18/climate-protestors-arrest-nottingham. See also H Porter, 'The Crushing of Eco-protest Brings Shame on Our Police' www.guardian.co.uk/commentisfree/2009/apr/19/police-powers-abuse-henry-porter.

[52] www.indymedia.org.uk/en/2009/04/427747.html.

[53] Ibid.

[54] 21 May 2010.

instead that adopted by HHJ Kemp in the non-climate change direct action case *R v Smith*[55] who, unlike HHJ Spencer, erred on the side of leaving the defences as matters for the jury. Nevertheless, the jury were not convinced by the defences and on 14 December 2010, the defendants were unanimously found guilty of the conspiracy offence. While the case thus failed to provide a softening repeat of Kingsnorth, repeating instead the failure of Drax, the second trial produced a more favourable outcome, and the events surrounding it cast doubt on the convictions secured in the first trial.

The six defendants at the second trial were those who had not made their minds up about whether to take part in the action. Their trial, at the start of 2011, collapsed in dramatic fashion after an undercover police officer who had infiltrated the protesters apparently 'turned' and offered to give evidence on behalf of the defence. Questions were then raised about the possible withholding of evidence by the police in the first trial, which led the Director of Public Prosecutions to order a review into the safety of the defendants' convictions.[56] Although not quite a Kingsnorth-type outcome, the questionable personal behaviour by the undercover police officer (including allegations of sexual relationships with activists)[57] combined with alleged agent provocateur behaviour[58] did arguably have the effect of softening the hard framing attempted by the police and prosecution.

Another police framing of climate change activists as extremists deserving of a robust policing response rather than a facilitative one more appropriate to peaceful protesters could be seen at the G20 protests in the City of London on 1 April 2009.[59] Although the G20 protests were not limited to climate change, climate change protesters formed a significant number at the event, with a special 'Fossil Fools Day' Climate Camp established outside the European Climate Exchange in Bishopsgate. Policing at the G20 was particularly aggressive, with the death of one innocent bystander[60] and numerous injuries to others, and the use of controversial 'kettling' or containment techniques preventing protesters from leaving. The police's

[55] Before Lewes Crown Court (sitting at Hove), 1 March 2010 (unreported, transcript on file with author). The case involved defendants charged with conspiracy to cause criminal damage at EDO in Brighton, which manufactured military equipment for Israel.

[56] O Bowcott, 'CPS Reviews Environmental Activists' Convictions' www.guardian.co.uk/environment/2011/jan/28/cps-reviews-environmental-activists-convictions.

[57] P Lewis and R Evans, 'Tricked, Betrayed, Violated: Did Police Spy Use Sex to Win Activists' Trust?' www.guardian.co.uk/environment/2011/jan/11/undercover-officer-sexual-tactics?intcmp=239.

[58] 'Lawyer Criticises Police on Undercover Pc Mark Kennedy', www.bbc.co.uk/news/uk-12152484.

[59] See Climate Camp Legal Team, 'Demonstrating Respect for Rights? The Policing of The Climate Camp in The City on 1 April 2009' (18 April 2009) 4–7, climatecamp.org.uk/get-involved/working-groups/legal/g20_report.pdf.

[60] Ian Tomlinson.

handling of this event—and its invasive policing of previous Climate Camps such as Kingsnorth in 2007—has been the subject of critical official reports[61] and also led to two sets of related judicial review proceedings.[62]

Though a causal link is difficult to establish, after this outpouring of criticism against the invasive policing of climate protest, we saw a return to a much more low-key, soft style by the police at climate events—evident at, for example, the 2009 Climate Camp at Blackheath in London.[63] However, whether such an approach will remain after the 2010 student protests against the Coalition Government's plans to introduce significant University tuition fees, remains to be seen. While the police employed a low-key approach at the first major London protest on 10 November 2010, the violence at Conservative party headquarters which took the police by surprise, arguably led to a much harder policing style being adopted at future student protest events in London in late November and December of the same year. Extensive attempts at kettling were made at both of these latter demonstrations.

A move away from hard to soft, more facilitative policing of protest is clearly supported on human rights grounds. Thus, in *Gillan and Quinton*,[64] the European Court of Human Rights (ECtHR) ruled (contrary to UK court judgments including the House of Lords)[65] that broad stop and search powers under section 44 of the Terrorism Act 2000 used against a protester and a journalist on their way to an arms fair in Docklands, London, were in breach of Article 8 of the Convention. The UK courts' support for hard policing styles will be further tested by the reaction of the ECtHR to the *Austin* case lodged before it,[66] in which the House of Lords had previously upheld the legality of kettling techniques employed against May

[61] See, eg Her Majesty's Inspectorate of Constabulary (HMIC), *Adapting to Protest* (July 2009); HMIC, *Adapting to Protest: Nurturing the British Model of Policing* (November 2009) and House of Commons Home Affairs Committee, 'Policing of the G20 Protests', 8th Report of Session 2008–09, HC 418, 29 June 2009. See also P Lewis, 'MPs to Examine G20 Police Tactics as New Claims Emerge' www.guardian.co.uk/politics/2009/apr/20/police-assault-g20-protests.

[62] One set of proceedings involves the policing of the 2009 G20 protest (*R (McClure and Moos) v Commissioner of Police for the Metropolis*—being heard in March 2011); the other involved policing of the 2008 Kingsnorth Climate Camp (*E, T and Morris v Chief Constable of Kent Police* [2009] EWHC 2264 (Admin).

[63] P Walker and P Lewis, 'Police Adopt Low Profile as 1,000 Climate Camp Activists Set Up Base' www.guardian.co.uk/environment/2009/aug/27/climate-camp-environment-activists-police.

[64] Application no 4158/05, 12 January 2010.

[65] *R (Gillan) v Commissioner of Police for the Metropolis* [2006] UKHL 12; [2006] 2 AC 307.

[66] P Lewis, 'Metropolitan Police's "Kettling" Tactic Challenged in European Court' www.guardian.co.uk/politics/2009/jul/19/metropolitan-police-kettling-human-rights; www.christiankhan.co.uk/ViewNews.asp?NewsID=171. Other applications to the ECtHR on kettling have also been made—eg *Lowenthal and O'Shea v UK*.

Day protesters in London in 2001.[67] A similar human rights support for a softening of hard policing can be seen in the report by the UK Parliament's Joint Committee on Human Rights, *Demonstrating Respect for Rights? A Human Rights Approach to Policing Protest*,[68] which recommended, inter alia, that counter-terrorism powers should not be used against peaceful protesters.[69]

The G20 protests also demonstrate that attempts at framing do not always work. Framing climate protesters as extremists or terrorists and policing them accordingly will only work if this frame 'resonates' with the wider public.[70] The 'massacre of the innocents' nature of the G20 protest in particular meant that the attempt by the police to apply a 'very hard' frame did not stick: given the soft behaviour of, in particular, the vast majority of Climate Camp protesters at the time,[71] it was difficult to see how it could have done.

CONCLUSIONS

There is a temptation with climate change litigation to concentrate solely on proactive litigation by the climate change movement and also, perhaps, by its sceptical counter-movement.[72] However, certainly in the UK, climate change litigation has arguably received more media attention in the shape of reactive litigation following on from highly visible protest actions. The Kingsnorth 6 are perhaps the best example of this. One of the aims of this chapter has been to re-orientate legal academic attention on climate change litigation in this form.

[67] *Austin v Commissioner of Police of the Metropolis* [2009] UKHL 5; [2009] 1 AC 564.

[68] Seventh Report of Session 2008–09, HL Paper 47-I/HC 320-I.

[69] I am grateful to Mike Schwarz for drawing the rights sources in this paragraph to my attention.

[70] On frame resonance, see, eg D Snow and R Benford, 'Master Frames and Cycles of Protest' in A Morris and C Mueller (eds), *Frontiers in Social Movement Theory* (Yale University Press, 1992) 133–55.

[71] Attested to in the initial HMIC Report, above n 61, 52.

[72] Eg *Dimmock v Secretary of State for Children, Schools and Families* [2007] EWHC 2288 (Admin); [2008] 1 All ER 367. The judicial review claim, brought by Stuart Dimmock, a lorry driver, parent and school governor, argued that there had been a breach of ss 406–07 of the Education Act 1996 because of the lack of political neutrality in showing Al Gore's film, An Inconvenient Truth, in schools in the UK. The High Court ruled that the film had to be accompanied by guidance in relation to elements where Gore's film was said to have overstated its case beyond the established majority science. The case can be regarded as a movement one insofar as it was supported by key figures in the UK climate change denial movement, including Viscount Monckton and quarry magnate Robert Durward (with the latter providing the financial backing)—see www.guardian.co.uk/uk/2007/oct/14/schools.film and J Hoggan and R Littlemore, *Climate Cover Up: the Crusade to Deny Global Warming* (Vancouver, BC, Greystone Books, 2009) 149.

In adopting a social movement perspective, the chapter has focused on tactics or strategies themselves—on the part of both the climate change movement and the police—as a form of framing. The UK climate change protests of the early 2000s can be seen as a contest between the police's attempt to apply a hard set of frames—ranging from criminality, through extremism, to terrorism—and the climate movement's efforts at soft counter-framing. This contest arguably reached a high water mark with, on the one hand, the Kingsnorth acquittals and, on the other, the events of the G20 protests of 2009. However, after that time, the failure of the Drax 22 to secure a Kingsnorth-type acquittal significantly dented the more heroic court-inscribed softening which that earlier case had managed to achieve. While the criticism attached to the G20 policing and the new emphasis on rights initially appeared likely to lead to a decline in the effective use of hard frames by the police on ordinary demonstrations, the 2010 UK student protests have made that more questionable. Although the Ratcliffe-on-Soar case may, in the circumstances, have been something of a draw in terms of court-based hard in contrast to soft frames, the June 2010 jury trial of the 'Climate 9' who had been protesting against the expansion of Aberdeen airport,[73] suggests that Kingsnorth may have been something of a one-off. The climate defence of necessity for their actions in shutting down the airport for two and a half hours, failed in the Scottish Sherriff's Court, where they were found guilty of breach of the peace by the jury. The planned appeal[74] should help to reveal whether Kingsnorth-style softening has a future in court. That said, one should beware of focusing solely on framing by courts and juries. While attempts at reframing the prosecution's hard frames of criminality with the soft frame of lawfulness may fail to resonate in court, the repetition of many such cases with the running of climate defences, also frames the issue for a wider, media audience. Though judges and juries may remain unconvinced, protesters may still hope that their soft 'lawful' framing will resonate with the wider public, triumphing over the criminality frame ultimately imposed on them by the courts.

[73] www.climate9.com/.

[74] Ibid.

4

Climate Change and Paradoxical Harm

ROB WHITE

INTRODUCTION

CLIMATE CHANGE AND global warming pose a number of important questions for humanity. Not the least of these are problems relating to security and the management of social conflict. For example, there is a range of existing and potential social conflicts surrounding global warming,[1] including conflicts over environmental resources such as water and food; conflicts linked to the climate-induced migration of peoples; conflicts over differential exploitation of resources (particularly evident in bio-piracy in relation to indigenous people); and conflicts over transference of harm in the form of, for example, cross-border pollution. All of these are in some way linked to global changes in climate.

As governments and communities search for solutions to the underlying issues of climate change, and adopt measures to mitigate and adapt to the consequences of global warming, other problems will inevitably emerge. Indeed, the aim of this chapter is to illustrate how certain responses to climate-related issues generate their own sort of negative feedback loop, resulting in further degradation of environments and additional threats to basic human rights. In their own way, the pressures associated with these emergent harms will add to the social conflicts already associated with climate change.

The chapter explores how particular solutions to the problems associated with climate change are, in turn, generating new forms of social and environmental harm. *Paradoxical harm* is harm that arises out of an apparent contradiction (for instance, we have to pollute certain parts of the planet

[1] R White, 'Climate Change and Social Conflict: Toward an Eco-Global Research Agenda' in K Kangaspunta and I Marshall (eds), *Eco-Crime and Justice: Essays on Environmental Crime* (Turin, Italy, United Nations Interregional Crime Research Institute [UNICRI, 2009)]; D Mwiturubani and J-A Wyk (eds), 'Climate Change and Natural Resources Conflicts in Africa (Pretoria, Institute for Security Studies, 2010).

in order to save it from other types of pollution). Specific examples of paradoxical harm include the adoption of compact fluorescent light globes to save energy (but which contain toxic mercury), promotion of nuclear energy (but which involves disposal of nuclear waste), and carbon emission storage (that penetrates and despoils the subterranean depths of land and sea). Such harm is paradoxical precisely because the harm stems from the pursuit of sectional social interests that inevitably fashion responses to, rather than resolution of, the key contradictions of the present age (namely, preservation of the capitalist growth economy in contrast to transformation toward sustainable ecology).

Paradoxical harm is not the same as unintended consequences. In many instances the (new) harms are known, and the acts leading to the generation of the harms is intentional. The harm is paradoxical in the sense that while seemingly contradictory (we generate harms as a means to forestall other harms), it is perfectly logical from the point of view of the imperatives of the system as a whole. Economic and social interventions that sustain the status quo (in favour of hegemonic nation-states and the leading transnational corporations, and that include maintaining the viability of 'dirty' industries) are favoured over those that might tackle the key drivers of climate change and that could diminish the burgeoning threats to ecological sustainability worldwide. Universal human interests are thus superseded by pursuit of specific sectional interests, to the detriment of all.[2] While this happens by design, there is no grand plan. It is an outcome of a global system of production and consumption that is fundamentally premised upon private profit and narrow self-interest. The triumph of neo-liberalism is simultaneously the death knell of collective well-being. This, too, is the lynchpin of contemporary class struggles occurring around the globe.

The problem of global warming is complex and entails many different factors and elements. It involves stratospheric ozone depletion, and deforestation, through to air pollution associated with urban life and certain forms of agricultural production.[3] So, too, responding to climate change and global warming entails many different types of human intervention. The specific concern of this chapter is to investigate government and business responses to climate change in order to assess how these are related to new or extended harms against humans, nonhuman animals and specific environments. The chapter discusses two broad areas in which paradoxical harm is evident: food production and energy production. As illustrated below, the policies and practices adopted to address these issues, while

[2] R White, 'Green Criminology and the Pursuit of Social and Ecological Justice' in P Beirne and N South (eds), *Issues in Green Criminology: Confronting Harms Against Environments, Humanity and Other Animals* (Collompton, Willan Publishing, 2007).

[3] United Nations Environment Programme, *Global Environment Outlook* (New York, UNEP, 2007).

giving the appearance of responding to climate change, nevertheless deliver highly problematic answers to global warming.

CLIMATE CHANGE, FOOD PRODUCTION AND BIODIVERSITY

Changes in climatic conditions are putting new and additional pressures on existing global food stocks. One response by governments and agribusiness has been to foster ever greater reliance upon large-scale agricultural techniques and methods, and on new technologies such as the use of genetically modified organisms (GMOs). This has involved converting land to industrial forms of agricultural production, and the application of practical restrictions on what is being grown and how. These, in turn, have implications for both climate change and human well-being.

The corporatisation of agriculture has been accompanied by significant changes in land use. For example, there has been a major expansion in cropland worldwide. 'Viewed in a wider historical context, more land was converted to cropland in the 30 years after 1950, than in the 150 years between 1700 and 1850'.[4] The environmental impact is loss of habitat and biodiversity; soil water retention and regulation; disturbance of biological cycle; increases in soil erosion, nutrient depletion, salinity, and eutrophication. For humans, there is greater exposure to agrochemicals in air, soil and water.

In addition to the corporatisation of agriculture, food production is increasingly influenced by localised changes in biodiversity, and by the overarching effect of global warming. Climate change is altering the physical and biological world in a number of ways. For instance, two-thirds of the North Sea's fish species have shifted in both latitude and depth.[5] This example signals a more general issue, namely, large-scale movement of plant and animal species that is invasive to the endemic or native population of particular geographical areas. Simultaneously, local species may be threatened by changes in temperature, moisture, wind and carbon dioxide that diminish their ability to withstand hitherto familiar pathogens, much less newly invading species. Moreover, as local climates and temperatures change, greater pressure is placed on local environments as local producers convert land to industrial uses (eg agriculture, forestry and pastureland) in response to phenomenon such as desertification.[6]

[4] Ibid, 86.

[5] BioSecurity New Zealand, 'Climate Change and Biosecurity', *Biosecurity Summit 2006* (Wellington, BioSecurity New Zealand, 2006).

[6] United Nations Development Programme, *Biodiversity Conservation and Sustainable Land Management*. Website information (UNDP, 2010) http://www.undp.org/biodiversity/.

Ironically, one of the greatest threats to biodiversity is the industrialisation of agriculture (incorporating the use of seed and other patents) since this is one of the greatest causes of erosion of plant genetic and species diversity. The use of genetic modification technologies in food production is arguably one of the most publicly recognised and feared responses to world hunger. One consequence of the industrialisation of agriculture, combined with and intensified by application of GMO technology, is that biodiversity is systematically reduced.[7] For example, it has been estimated that 75 per cent of crop diversity has been lost over the past century.[8] In other words, there is a trend toward monoculture, since uniformity means ease of cultivation and harvest, and higher yield, which translates into higher profit.

This simplification of production generates paradoxical harms for biodiversity and potentially threatens future food production.

> One consequence of the erosion of plant genetic diversity is that the capacity of the economically preferred plants to resist pests and diseases is compromised. The marketability of plant produce is not necessarily coterminous with the inherent superior quality of the plants to be marketed or selected for mono-cropping. Given the potential utility of plants that market forces may erroneously dismiss as economically useless, the short-sighted depletion of the plant genetic pool can be both costly and dramatic.[9]

Put simply, 'over the ages farmers have relied upon diverse crop varieties as protection from pests, blights and other forms of crop failure'.[10] Reducing this diversity affects the inbuilt mechanisms that have historically protected the soil and vitality of the overall agricultural process. Moreover, intensive use of land and soils that rely upon chemical additives to ensure productivity, rather than for example traditional methods of crop rotation, further diminishes longer term agricultural viability.

The reduction of biodiversity itself feeds into the larger pattern of climate change.[11] This occurs through adoption of methods of food production that lead to increases in carbon emissions. Consider for example food production in New Zealand, which is heavily dependent on international trade, with exports contributing 29 per cent of Gross Domestic Product and natural resource-based exports (from agriculture, forestry, fishing and aquaculture) accounting for a large share.[12] Indeed, the relatively recent

[7] M Pollen, 'Unhappy Meals' *New York Times Magazine*, 28 January 2007, 38–47, 65–70.

[8] H French, *Vanishing Borders: Protecting the Planet in the Age of Globalization* (New York, WW Norton & Company, 2000) 61.

[9] I Mgbeoji, *Global Biopiracy: Patents Plants, and Indigenous Knowledge* (Vancouver, UBC Press, 2006) 181.

[10] French, above n 8, (2000) 81.

[11] See UNEP, above n 3.

[12] Organization for Economic Cooperation and Development, 'Conclusions and Recommendations', *OECD Environmental Performance Review of New Zealand* (Paris, OECD, 2007) 1.

history of New Zealand (the past two hundred years) is one of remarkable transformations in land use. While 80 per cent of Aotearoa New Zealand was once forested, only about 20 per cent of indigenous forest-cover now remains—that is, only 22 per cent of the land surface of the indigenous habitat remains in more or less primary condition.[13] Moreover, 'Land use analysis shows a net loss of nearly 175 km^2 of indigenous habitat between 1996 and 2002'.[14] In many cases this land clearance has occurred without proper permits and resource consents.

Transformations in land use are directly linked to export earnings. This emphasis and reliance upon certain types of export earnings has a number of consequences. For instance, pollution and degradation is directly linked to the economically productive use of land. Such issues are particularly central to discussion of environmental harm in the New Zealand context. They are associated with both past practices and contemporary realities. For example, just over 39 per cent of New Zealand's total land cover is pasture, and it is from within the pastoral industry that we see the main environmental problems surfacing.

The New Zealand Ministry for the Environment has observed that

> The amount of nitrogen fertiliser used in New Zealand has increased about ten-fold since 1985 and had doubled since the mid-1990s. Nitrogen from livestock manure, which contributes around five times the amount of nitrogen to the land as nitrogenous fertilisers, also steadily increased. These changes coincide with the trend towards more intensive forms of farming; particularly dairy farming, with its high density of grazing stock. Dairy cows excrete almost seven times the amount of nitrogen and phosphorus in their faeces and urine as breeding ewes, and around three-and-a-half times that of breeding hinds (deer).[15]

The combination of the sheer number of animals and the use of nitrogen fertilisers is having a major negative impact on the surrounding environments. Indeed, dairy farming has been identified as the single largest cause of environmental decline in New Zealand, due to fertiliser and animal waste runoff from farms, and the use of water itself.[16] This is placing severe demands on New Zealand's natural resources, some of which are being used unsustainably. Irrigation in arid regions, cow dung and urine that affects nitrates in groundwater, and sewerage discharge into freshwater lakes are all issues of concern, as noted in an OECD report on environmental performance.[17] The same report notes that 'In contrast to many OECD

[13] Conservation International, 'New Zealand: Human Impacts' (2007) www.biodiversity-hotspots.org/xp/Hotspots/new_zealand/Pages/impacts.aspx.

[14] OECD, above n 12, 5.

[15] New Zealand Ministry for the Environment *Environment New Zealand 2007: Summary* (Wellington, Ministry for the Environment, 2007) 42.

[16] Scoop Independent News, 'Professor slams environmental performance report' (2008) [www.scoop.co.nz/stories/SC0802/S00062.htm].

[17] OECD, above n 12.

countries, GHG [green house gas] emissions from agriculture (eg, methane and nitrous oxide) account for some 50% of the national total, and are rising. Changes in agricultural production have led to increased intensity of inputs, including fertiliser and irrigation water, with consequent increases in environmental pressures'.[18]

Things are hardly likely to get better vis-à-vis good environmental practices and outcomes in the near future. This is so for several reasons. First, there is a huge export market for agricultural and pastoral commodities. The global situation in regards to grain, for example, and more generally the scarcities associated with certain foods, means that profit is to be made and farmers (and their political fellow travellers) will be keen to secure as much return as possible in such circumstances. Intensification of production, that also includes the extensive use of fertilisers, is one by-product of the global pressures on food production. As the worldwide competition over resources intensifies, so too will those forms of food production that most readily contribute to global warming.

Secondly, there is the related issue of how to maintain resilient ecosystems in the face of pressures associated with climate change and inappropriate land uses by humans. 'Biodiversity brings resilience to ecosystems by spreading risk and making ecosystems more able to reorganise and adapt after change and disturbance. Ecosystems are particularly resilient if there are many species contributing to the same service'.[19] Responding to climate change involves efforts to protect and maintain bio-diversity. Thus, feeding the world through forms of production that reduce biodiversity is, ultimately, counterproductive and a generator of social and environmental harm.

CLIMATE CHANGE, ENERGY PRODUCTION AND BIOFUELS

Paradoxical harms stemming from climate-related energy issues can be analysed in terms of the use of alternative energy sources, and efforts aimed at dealing with carbon emissions. In each case the answer to the energy crisis involves measures that in some way contribute to other forms of environmental harm. The consequences for humans, local environments and nonhuman species are in some cases unknown, and certainly demand consideration of the precautionary principle.

Much public debate has occurred over the regulation and reduction of carbon emissions. At the heart of the matter is the fact that carbon emissions directly contribute to global warming, and that without adequate strategies of mitigation and adaptation the problems associated with climate change

[18] Ibid 7.

[19] Victorian Government, *Securing Our Natural Future: A White Paper for Land and Biodiversity at a Time of Climate Change* (Melbourne, Victoria, 2009).

will get worse before they get better. The urgency surrounding the reigning in of carbon emissions has been matched by the audacity of businesses in lobbying to defend their specific economic interests.

Given that key proposed solutions to global warming include carbon emission trading, and the sequestration of carbon emissions (in the form of carbon emission storage, as well as through protection of the world's tropical and other old growth forests), the introduction of a wide range of regulatory sanctions is inevitable. One issue of concern is the tendency for many businesses to either publicly pronounce their green credentials with little evidence, or to exaggerate the potential threats to their industry or business as a result of specific government policies (such as carbon reduction schemes). In either case, there is an element of green washing and basic dishonesty. Such corporate conduct is at best misleading, and at worst may constitute a form of fraud. Responding to climate change in this instance revolves around the sentiment and public discourses of businesses, including unscrupulously portraying their interests in ways that, one way or another, undermine a stronger regulatory environment vis-à-vis carbon emissions.

The influence of specific business interests is reflected politically in government responses to climate change in relation to energy production. For example, the push toward biofuel production reflects the interests of large agricultural businesses, who patent the monocultural crops designed as 'energy crops'. Powerful sectoral interests, including car manufacturers and grain farmers, have benefited from the search for energy alternatives to fossil fuels. The shift to biofuel is seen as a key source of green fuel supply for the world's car manufacturers. Greater demand for biofuel crops such as corn, palm oil or soya also means that farmers are finding the growing of such crops very lucrative economically. The hybrid car is the flavour of the month for all concerned.

This trend toward biofuel production is generating its own paradoxical harms.[20] First, the use of crops for fuel is leading to food price rises and food shortages.[21] This became evident when Mexicans protested in the streets about the price of corn flour, a primary ingredient of tortillas. The situation was brought about by US corn growers selling an increasing portion of their harvest for manufacture of corn-based ethanol (a diesel-type fuel made from plants). Less corn for food equals higher food prices.

[20] Reliable Plant, 'New Study Favors Tree Over Corn as Biofuel Source' (Nov/Dec 2007) [www.en.engormix.com/forums_view.aspx?id=3952]; Reuters (21 January 2008), 'Committee calls for biofuel moratorium' [www.uk.reuters.com/article/2008/01/21/uk-biofuels-britain-idUKL1885287820080121]; News.scotsman.com (21 January 2008), 'Burning issue: Biofuel targets and subsidies: is it time for a moratorium?' available at, www.scotsman.com/news/burning_issue_biofuel_targets_and_subsidies_is_it_time_for_a_moratorium_1_10746.

[21] D Mitchell, 2008, '*A Note on Rising Food Prices*', Draft World Bank paper, circulated on-line by The Guardian Newspaper, website (guardian.co.uk/environment).

Secondly, the profitability of biofuel production is spearheading the establishment of large-scale plantations in places such as Indonesia and Brazil. This process has seen the clearing of rainforests and in some instances the forcing of Indigenous people off their lands. This deforestation process has been going on for a number of years, supported by organisations such as the International Monetary Fund. Clearing of land for export-oriented cash crops has been touted as a key strategy to lift developing countries' economic performance.[22] Biofuels provide yet another avenue to accelerate this process. Cutting down trees also, of course, has a direct bearing on global warming and regional climate change (eg less rain).

Thirdly, there is evidence that the nitrogen-based fertiliser used in corn production is itself causing environmental harm. Millions of kilograms/pounds of nitrates end up in the Gulf of Mexico each year, where they are causing a massive algae bloom. This bloom impacts negatively on the ecology of the Gulf: 'When the algae dies it sinks to the bottom, where it absorbs oxygen as it decays. In recent years that oxygen depletion has created an aquatic "dead zone" covering about 8,000 square miles in which shrimp, fish, oysters and crabs cannot survive'.[23]

Biofuel production in places such as the United States and the European Union is encouraged through strong incentives (eg tax credits) and mandates such as energy legislation (eg mandatory blending requirements). As indicated, the advent of biofuels has helped to push up global grain prices, and to bolster the prospects of grain-producing countries. However, it has also been accompanied by ecological costs in the form of degraded environments and social costs in the form of high prices for food, especially in less developed and import-dependent countries.[24]

According to some commentators,[25] the search for beneficial biofuels ought to focus on feedstocks that:

— Do not compete with food crops
— Do not lead to land clearing
— Offer real greenhouse-gas reductions
— Maximise social benefits

In practice, land, water and energy resources have effectively been put to the goal of producing bio-fuels rather than production of food for human

[22] See French, above n 8.

[23] Reliable Plant, above n 20.

[24] G Roberts, 2008, 'The Bad Oil on Ethanol: Biofuels Are Losing Favour but Some Governments Are Still Backing Them', *The Weekend Australian*, 31 May–1 June 2008, 20 Inquirer.

[25] L Rist, J Ser Huay Lee and L Pin Koh, 'Biofuels: Social Benefits' (2009) 326 *Science Letters* 1344; D Tilman, R Socolow, J Foley, J Hill, E Larson, L Lynd, S Pacala, J Reilly, T Searchinger, C Somerville and R Williams, 'Beneficial Biofuels—the Food, Energy, and Environment Trilemma' (2009) 325 *Science* 270–71.

consumption[26] Importantly, while it may seem beneficial to use renewable plant materials for biofuel and the use of crop residues and other biomass, food and biofuels are dependent upon the same resources for production. Given the problems of biofuel production (eg the release of large quantities of carbon dioxide associated with the planting and processing of plant materials for biofuels), and given that nearly 60 per cent of humans in the world are currently malnourished, the blunt conclusion is that 'There is simply not enough land, water, and energy to produce biofuels'.[27] Biofuels are thus not quite the panacea to environmental and energy crises that some supporters claim.

The notion that 'business can profit by protecting the environment'[28] is frequently linked to the idea of technological fixes vis-à-vis climate change, such as the development of hybrid cars, more technologically efficient appliances and lighting sources such as fluorescent bulbs. Burying the problem is also touted as a solution—whether this be radioactive waste or carbon pollution.

For example, uranium is not a renewable energy source, yet nuclear power is high on the list of preferred energy sources in the era of global warming. But nuclear energy also means nuclear waste. High level waste (HLW) is so radioactive that it generates heat and corrodes all containers. It must be stored above ground for 50 years so it can cool before being transported and disposed of. Intermediate level waste (ILW) arises mainly from the reprocessing of spent fuel and from general operations and maintenance at nuclear sites. It is typically packaged by encapsulation in cement in highly-engineered stainless steel drums or in high capacity steel or concrete boxes.

In the UK alone, 365 000 cubic metres of high and intermediate-level radioactive waste will soon have been accumulated from its existing post-war nuclear programme.[29] The decision in 2009 by the US government to end development of the Yucca Flat, Nevada repository for permanent disposal of radioactive waste has also highlighted waste issues in that nation. As of 2009, the United States has been left with some 60 thousand metric tons of spent reactor fuel in need of a permanent storage facility, but with no viable facility on the drawing board, and no indication of when—if ever—the situation will be resolved.[30] Problems in developing a suitable response relate to

[26] D Pimental, A Marklein, M Toth, M Karpoff, G Paul, R McCormack, J Kyriazis and T Krueger, 'Food Versus Biofuels: Environmental and Economic Costs' (2009) 37 (1) *Human Ecology* 1–12.

[27] Ibid, 9.

[28] H Baer and M Singer, *Global Warming and the Political Economy of Health: Emerging Crises and Systemic Solutions* (Walnut Creek, CA, Left Coast Press, 2009) 181.

[29] E Clarke, 2009, 'The Truth About ... Nuclear Waste', ClimateChangeCorp.com [www.climatechangecorp.com/content.asp?ContentID=6024].

[30] W Pickard, 'Finessing the Fuel: Revisiting the Challenge of Radioactive Waste Disposal' (2010) 38 *Energy Policy* 709–14.

technical and scientific issues, the Nimby [Not In My Back Yard] effect, and who is or ought to carry the financial burden over time.

Nuclear waste in the US is currently stored at 121 temporary sites in 39 states across the country.[31] Yet, 'Sixty years into the nuclear era there is no universally agreed upon strategy for the disposal of high level nuclear waste. Significantly more is being produced each year by nuclear powered electricity generation. But the power so generated is needed.'[32] At present, the only reasonable course of action, and moral obligation, is to bury this type of waste deep underground, rather than leaving the waste on the surface for an indefinite period—where each day might bring an incident that disperses it into the biosphere.[33]

The biggest challenge for governments is to find a site that is both geologically secure and acceptable to the local communities. This is compounded by the problem of capacity relative to need. In South Korea, for example, the large increase in projected nuclear power will inevitably accelerate the accumulation of spent fuel—but if the direct deposit option is pursued, South Korea would not be able to secure enough suitable sites for disposal, given its geographic profile.[34] Meanwhile, most countries are still in the planning stages for HLW repositories. In Japan, for example, the government has not yet started work on its HLW repository, even though the first experimental reactor was initiated in 1963.[35]

Then there is the effect of global warming on proposed storage strategies. For example, the near-surface facility and geological repository are both likely to be affected by the consequences of global warming—the landscape and hydrogeological regime at and around a disposal facility may change, as might the biosphere receptors, and the animal and human habitats. When considering the long-term evolution of the disposal system, major climate change (eg future glacial periods) should be assessed.[36] Global warming will have a major influence over short and long periods of time, as in the case of changing water infiltration through the multi-layer cover of a waste disposal site.

The environmental crisis stemming from climate change is a major impetus for a huge and rapid expansion of the nuclear power industry. The criminal and safety implications of these combined issues include illegal disposal of radioactive waste; unsafe containment of existing and immediate future

[31] Clarke, above n 29.

[32] Pickard, above n 30, 713.

[33] Ibid.

[34] W Ko and E Kwon, 'Implications of the New National Energy Basic Plan for Nuclear Waste Management in Korea' (2009) 37 *Energy Policy* 3484–88.

[35] M Zhang, M Takeda, H Nakajima, M Sasada, K Tsukimura and Y Watanabe, 'Nuclear Energy and the Management of High-Level Radioactive Waste in Japan' (2009) 14 (11) *Journal of Hydrologic Engineering* 1208–13.

[36] M Van Geet, M De Craen, D Mallants, I Wemaere, L Wouters and W Cool, 'How to Treat Climate Evolution in the Assessment of the Long-Term Safety of Disposal Facilities for Radioactive Waste: Examples From Belgium' (2009) 5 *Climate of The Past Discussions* 463–94.

waste; and externalisation of costs and wastes from core to peripheral countries. There is the accusation, for example, that French nuclear waste gets shipped to northern Africa and gets dumped in the desert sands.[37] This allows for a clean, green image within France to be maintained, because the real problem has been exported away. The demand for new forms of energy, within a context of reducing carbon emissions, may well open the door to new types of dumping and new transferences of toxic harm worldwide.

The replacement of incandescent light globes with compact fluorescent light globes (CFLs) also poses dilemmas for disposal. Mercury is the most toxic pollutant after radioactive substances on the Environmental Protection Agency list of industrial pollutants. One type of light globe is being replaced with another in order to save energy (climate change agenda), but the new CFL globe contains mercury and there are already serious questions about the mass waste disposal of such globes (pollution agenda).

For example, compact fluorescent lamps, used mostly in homes, are required under a new Australian standard to have a maximum of 5mg of mercury. This is seen by some as a major problem. For instance, 99 per cent of used fluorescent tubes and High Intensity Discharge (HID) lamps, which contain mercury, are currently dumped in public landfill sites, causing a serious and ever-increasing mercury pollution problem. The CEO of the Australian Council of Recylers, Ms Anne Prince, says that the move to fluorescent tubes without corresponding legislation governing their disposal is an ecological disaster in the making: 'We need to be smart enough to avoid creating a mercury pollution problem in order to fix a carbon pollution problem—we need another system, and fast'[38] Governments are in the process of consulting with various stakeholders over how best to respond to this new toxic waste stream.

Three disposal scenarios currently exist for CFLs—recycling, incineration, and landfilling. The issues that arise in the disposal of CFLs are that:[39]

— waste collectors and processors may be exposed to mercury as CFLs enter the waste stream, and that this exposure is likely to increase as more CFLs enter the waste stream
— mercury from lamps in landfills can be converted to methyl mercury. Methyl mercury is more toxic than elemental mercury and, when emitted to air, may be a risk to landfill workers

[37] F Bridgland (2006), 'Europe's New Dumping Ground: Fred Bridgland Reports on How the West's Toxic Waste Is Poisoning Africa' *Sunday Herald*, 1 October 2006, accessed 2/10/2006.

[38] Australian Council of Recyclers, 'Ecological Disaster Forecast Unless Fluorescent Tubes Recycled' *Media Release*, 27 March 2007.

[39] Equipment Energy Efficiency Committee, *Regulatory Impact Statement for Decision: Proposed MEPS [Mini-Energy Performance Standard] for Incandescent Lamps, Compact Fluorescent Lamps and Voltage Converters* (Canberra, Department of the Environment, Water, Heritage and the Arts, 2009).

— if CFLs are processed in Alternative Waste Treatment facilities the mercury they contain may contaminate compost and render it unusable
— mercury can also escape from landfills into the environment or into ground water as leachate
— mercury escaping from landfills in various forms can contribute to overall mercury pollution that can bioaccumulate in the environment and affect human health and animal health.

Recycling is seen as one of the best solutions to this, but this, too, raises issues of where the waste is to be recycled, under what conditions and by whom.[40] In some cases recycling is simply used as a cover for the transfer of toxic waste to developing countries.[41]

CONCLUSIONS

Developments relating to food production and energy production reveal a series of paradoxical harms that are generated in the context of strategic decisions governing how production (and consumption) take place. Figure 1 provides a summary and further examples of this.

Critical analysis of key areas such as food and energy alerts us to the ways in which sectional interests (that is, particular transnational corporations such as those pertaining to agribusiness and the nuclear industry, and particular hegemonic nation-states such as the US and China) are vying for control over and exploitation of resources vital to human existence. Concepts such as human rights, ecological citizenship and the global commons can be developed in ways that assert the primacy of 'climate justice' over these kinds of narrow sectional interests (see for example, Green and Ward in relation to use of human rights criteria as a means to deal concretely with state crime).[42] The causes and the effects of climate change can be analysed in terms of global distributions of social power, and the winners and losers in the global warming stakes. Current mitigation and adaptation strategies are in this respect profoundly unjust.

[40] Basel Action Network/Silicon Valley Toxics Coalition, *Exporting Harm: the High-Tech Trashing of Asia* (Seattle and San Jose, BAN/SVTC, 2002); R White, *Crimes Against Nature: Environmental Criminology and Ecological Justice* (Collompton, Willan Publishing, 2008).

[41] J Clapp, 'Seeping Through the Regulatory Cracks' (2002) XXII *SAIS Review* 141–55.

[42] P Green and T Ward, 'State Crime, Human Rights, and the Limits of Criminology' (2000) 27 (1) *Social Justice* 101–15.

Area of concern	Key issues	Policy and Practice Interventions	Paradoxical Harms	Examples	Outcomes
Food	Shortages Global unequal distribution Social conflict	Subsidies Land conversion Patent protections Market advantage	More of certain types of food = less biodiversity More of certain types of food production = increased global warming	Industrial forms of agricultural and pastoral production Adoption of bio-technologies	Pollution and contribution to green house gases Loss of habitat and biodiversity
Energy	Alternative sources needed Global unequal distribution Economic growth associated with consumption and energy use	Adoption of energy sources that emit less carbon Transfer of production of emissions to periphery countries Recycling	Alternative energy uses = new waste streams Alternative energy sources = increased carbon emissions	Poor regulation of carbon emissions Biofuels and deforestation Greater reliance on nuclear power New compact fluorescent light globes	International transfer of problem to poor countries Replacement of food crops with energy crops, and global warming Radioactive waste disposal, additional forms of toxic pollution

Figure 1. Food and Energy Production and Paradoxical Harm

5

Corporate Governance and Climate Change

SALLY WHEELER

INTRODUCTION

THE WORLD'S RESOURCES are finite in respect of fossil fuels and many minerals and, when resources such as fertile soil and clean water[1] are added, unevenly distributed in terms of location, access and use. Attempts to extract oil, coal and other minerals as cheaply and as quickly as possible by the extraction corporations[2] have added to water shortages and water pollution as well as altering terminally some landscapes and increasing emissions more generally. The control over land use exerted by agri-business, in efforts to satisfy the demand for cheap food from an ever-increasing world population and the demand for meat[3] from world population increasing in size and affluence, results in environmental degradation as water sources are polluted or exhausted and soil is thinned and contaminated by chemicals and eroded by irrigation. Bio-diversity, important for species survival, is threatened by the cultivation of particular, sometimes genetically modified, high resistance crops.[4] It is the often the case that those who benefit the least from these activities suffer the most from their adverse effects.[5] Corporations operating in these sectors are easy targets for the various environmental lobby groups. However nearly

[1] C Vösömarty, 'Global Threats to Human Water Security and River Biodiversity' (2010) 467 *Nature* 555.

[2] Extraction corporations are the largest sector presence on the FTSE 100.

[3] Not only does animal rearing require large amounts of land for grazing, feed growing, considerable water supplies to deal with feeding and waste but cattle in particular emit methane into the atmosphere. Meat production is responsible for a greater amount of greenhouse gas than the transport sector, *Livestock's Long Shadow* (Food and Agriculture Organization, Rome, 2007).

[4] B Johansen, *Global Warming in the 21st Century, vol 3: Plants and Animals in Peril* (Westport, CT, Praeger, 2006).

[5] JT Roberts, P Grimes and J Manale, 'Social Roots of Global Environmental Change: A World Systems Analysis of Carbon Dioxide Emissions' (2003) 9 *Journal of World Systems Research* 277, 288.

all corporate activity occurs on a spectrum that involves the conversion of natural resources into a finished consumer product through a supply chain or the financing of such a conversion and thus environmental effects that exacerbate climate change are involved.[6] Corporations are ideally placed for a leading role in preserving resources, maintaining ecosystems, discovering and fostering technology and resource renewal strategies if, and only if, the appropriate drivers for this behaviour are in place.[7] The financiers of corporate activity are also ideally placed to punish corporations with poor environmental records and to reward those that engage with the struggle to adapt to climate change.[8]

Global structures of investment capital mean that there are numerous intersections between the corporate economy and climate change. Here I am concerned with corporate governance and not with corporate compliance stra tegies, corporate disclosure initiatives or corporate lobbying strategies in relation to regulators. I have taken corporate governance, for this chapter, to mean the opportunities that institutional investors have to shape corporate strategy in their investee companies and the means by which this might occur. The first section of the chapter looks at the landscape of investment. The second and longer section of the chapter looks at the rise of 'responsible ownership' and what this might mean in relation to climate change. It also includes a discussion of shareholder resolution based activism. I do not, for reasons of focus and length examine the recent reforms to the proxy machinery in the US[9] and the debates on shareholder democracy that are current there.[10] Nor do I deal with, in any detail, the problems of collective action and free riders in the context of shareholder activism.

Section 1

DIRECTING STRATEGY IN THE CORPORATION

To expound on the links between corporate governance and climate change, we need first to identify the power blocks and relationships within the

[6] Richardson describes corporations that are financiers in these circumstances as 'unseen polluters'. His work is concerned with the potential for a more creative regulatory approach to financial sector corporations in particular through the use of legal standards for social responsible investment, see B Richardson, 'Putting Ethics Into Environmental Law: Fiduciary Duties for Ethical Investment' (2008) 46 *Osgoode Hall Law Journal* 243.

[7] P Shrivastava, 'The Role of Corporations in Achieving Ecological Sustainability' (1995) 20 *Academy of Management Review* 936, 940.

[8] B Richardson, 'Financing Environmental Sustainability' in S Benn and D Dunphy (eds), *Corporate Governance and Sustainability* (Abingdon, Routledge, 2007) 122.

[9] Dodd–Frank Wall Street Reform and Consumer Protection Act and the rules enacted by the Securities and Exchange Commission in August 2010 under this legislation on proxy access.

[10] H Manne, 'The 'Corporate Democracy' Oxymoron' (2 January 2007) *The Wall Street Journal*.

corporation. The strategic direction of any corporation is determined by the board of directors who are informed in this by the management structure that they have put in place. There is some dissonance between the Companies Acts and common law on the one hand and the UK Corporate Governance Code[11] on the other as to the role of shareholders in the corporation. The provisions of this Code are not enforceable by shareholders or any other party or group but a statement of the extent to which there has been compliance with the Code by the corporation in question and the reasons for non-compliance are a requirement of continued stock exchange listing. The market then values this information and the value can be determined by looking at stock price movements. From reading the legislative framework for the structure of the corporation one could be forgiven for thinking that shareholders were comprised of disconnected individuals who were at best sometimes thrown together by sharing a minority interest, which then attracted certain rights of standing against the corporation in limited circumstances,[12] and that the shareholders' meeting where majority rule prevails, whether the Annual General Meeting or a shareholder called meeting[13] was the fulcrum of governance.[14] Shareholders have the right to appoint and remove directors,[15] change the corporate constitution[16] and decide whether to issue shares.[17] They have limited rights to dictate to directors[18] or to restrain directors from particular courses of action, as one would expect given the professional nature of corporate management and directorship, and so presumably would sell their shares rather than engage with the corporation on matters of strategic direction.

An examination of the UK Code of Corporate Governance creates a rather different picture of the levers that shape governance within the corporation. While the annual general meeting is recognised by the Code as a significant event for shareholders (principle E2 and supporting principles), it is clear that shareholder influence over corporate strategy is to come from a continuing dialogue between the board, through its chair and senior independent director, and *major* shareholders (Code provision E1.1). There is an assumption within the Code that there are two types of shareholder; large and small. Large or major shareholders are given an enhanced position within the corporation's governance structure that enables them to comment

[11] Previously known as the Combined Code on Corporate Governance, a new revised UK Code of Corporate Governance was issued by the Financial Reporting Council on 28 May 2010.

[12] CA 2006 s260-269, s994(1).

[13] CA 2006 s303.

[14] CA 2006 s319A.

[15] CA 2006 s168.

[16] CA 2006 s21.

[17] CA 2006 s549.

[18] See *Automatic Self-Cleansing Filter v Cunninghame* [1906] 2 Ch 34 and CA 2006 Table A, reg 4; J Hill, 'The Rising Tension Between Shareholder and Director Power in the Common Law World' (2010) 18 *Corporate Governance* 344.

The decline of individual participation in the equity markets is mirrored by the rise in institutional investment. Institutional investors manage the financial participation efforts of millions of individuals. Pension funds for both public and private sector employers represent the retirement savings of those of individuals. The rapid conversion of final salary schemes into defined contribution schemes across the world,[36] the rise of sectors of employment where occupational pension scheme provision was not present,[37] declarations that the pension system is unsustainable for the future without increased individual contributions[38] and in the UK a specific policy of the Thatcher era of encouraging individuals to opt out of state-based provision[39] and into a plan of their own choice has pushed many individuals into market-based financial participation through the use of collective investment vehicles such as ISAs and investment trusts, often made more attractive by tax incentives created to boost a culture of saving for retirement

INSTITUTIONAL INVESTORS, FINANCIALISATION AND SHAREHOLDER VALUE

The 1980s saw a sustained period of deregulation regarding the type of investments that could be offered as structured financial products[40] on

[36] P Bridgen and T Meyer, 'When do Benevolent Capitalists Change Their Mind? Explaining the Retrenchment of Defined-Benefit Pensions in Britain?' (2005) 39 *Social Policy and Administration* 764.

[37] P Ring, 'Security in Pension Provision: A Critical Analysis of UK Government Policy' (2005) *Journal of Social Policy* 343.

[38] See 'A New Contract for Welfare: Partnership in Pensions' Cm 4179 (London, HMSO 1998) and 'Security in Retirement: Towards a New Pension System' Cm 6841 (London, DWP, TSO, May 2006).

[39] What was contracted out of was SERPS or State Earnings-Related Pension Scheme which had been created to level out the difference between those who were members of an occupational pension scheme and those who were not. The Social Security Act 1985, the Pension Act of 1986 and the Financial Services Act of 1986 contained provisions that scythed through SERPS. The emphasis was on the establishment of personal pensions on a defined contribution basis facilitated by NIC rebates. Employers were no longer able to make it a condition of employment that individuals were members of their occupational scheme. Within ten years 68% of employees had left the SERPS scheme into contracted out provision. More significantly some 500,000 employees had left their occupational schemes and had taken out a personal pension. See N Deakin, *The Politics of Welfare* (London, Methuen, 1987) 138–54; and A Budd and N Campbell 'The Roles of the Public and Private Sectors in the UK Pensions System' in M Feldstein (ed), *Privatising Social Security* (Chicago, Chicago University Press, 1998) 99, 100.

[40] Structured financial products are products that seek to provide a particular need, eg income in ten years time. They work by changing the ratio of the cost of capital to risk so risk is diversified across a pool of investors or it is transferred for a fee to someone who is prepared to carry it. See J Bethel and A Ferrell 'Policy Issues Raised by Structured Products' in Y Fuchita and R Litan (eds), *New Financial Instruments and Institutions* (Washington DC, Brookings Institution, 2007) 167, 171–74.

both a retail and commercial basis.[41] This supply of financial products such as corporate bonds, life assurance policies and annuities[42] met both commercial demand and retail demand in the form of those individuals who were hoping to accumulate retirement wealth to replace the steady erosion of such provision within welfare states; those who wished to access higher returns than were available through more traditional 'high street' based mutual societies;[43] and pension funds which were eager to make up for the low returns of the 1970s, the fiscal demands created by the increased life expectancy of beneficiaries and the effects of the pension 'holidays' taken during the 1990s.[44] The pursuit of corporate profit growth through financial trading and the sale of financial instruments have dominated the corporate economy in recent years in a trend that is termed financialisation.[45] For example a glance at the business section of a UK Sunday newspaper reveals that the AA (formerly known as a road side car repair business and car insurance provider) now offers personal savings products as do Tesco and John Lewis (formerly renowned only as general retailers). The fruits of production and increasingly the methods of production can be become irrelevant or at least subservient[46] in the context of financial accumulation. Concerns for climate change and the articulation of those concerns within corporate governance are not part of the innovations which underscore financialisation of the corporate economy.

The desires and needs of institutional investors are paramount within the corporate governance paradigm; nowhere is that better illustrated than in the symbiotic relationship between executives and institutional investors that occurs in the pursuit of 'shareholder value'. 'Shareholder value' became the operational credo of many corporations during the 1990s although its roots have been identified as occurring earlier.[47] According to this model, corporate governance as a process existed solely to ensure that managers maximised the returns in the form of dividends and capital growth to shareholders.

[41] This period which was mirrored around the world in similar deregulation drives was called the Big Bang in the UK, see generally P Augar, *The Death of Gentlemanly Capitalism* (London, Penguin Books, 2000).

[42] 'In 2007 banks invested some £50 billion in manufacturing. In the same year they invested close to £800 billion in a variety of complex financial products.' *Britain's Broken Economy* (Lawrence and Wishart, The New Political Economy Network) 25.

[43] P Langley, 'The Making of Investor Subjects in Anglo-American Pensions' (2006) 24 *Environment and Planning* 919.

[44] P Langley, 'In the Eye of the 'Perfect Storm': The Final Salary Pensions Crisis and Financialisation of Anglo-American Capitalism' (2004) *New Political Economy* 539.

[45] G Krippner, 'The Financialization of the American Economy' (2005) 3 *Socioeconomic Review* 173, 181; and more generally I Erturk, J Froud, S Johal, A Leaver and K Williams (eds), *Financialisation at Work* (London, Routledge, 2008).

[46] On the schizophrenic nature of theories of value see M Sagoff, 'At the Shrine of Our Lady of Fatima or Why Political Questions are Not All Economic' (1981) 23 *Arizona Law Review* 1283.

[47] P Ireland, 'Financialization and Corporate Governance' (2009) 60 *Northern Ireland Legal Quarterly* 1.

This was a change from the portfolio-management through diversification that large corporations had practised before. The attraction that the shareholder value model held for executives was in the compensation packages they received—they began to be remunerated in stock options. Thus they became part of the shareholder value experiment, able to sell their shares at a far greater price than the value options were granted at.

The exact role that the pursuit of shareholder value played in the regulatory failures in the US in the early years of this century[48] and in the global financial crisis of 2008[49] is still a matter of debate but, in any event, it is not the focus of this contribution. What is clear, however, is that the crisis involved institutional investors, corporate executives and a cast of supporting characters such as ratings agencies, fund managers and securities analysts in the relentless pursuit of profit that has, in many cases, ignored other claims for the attention of corporate executives including that of the environment. Institutional investors, or asset managers acting on their behalf, have pursued policies of 'short-termism'[50] with behavioural traits such as herding[51] thrown in. This means that interests in corporations have frequently been traded rather than held so that profit is realised in order to contribute to quarterly performance figures and so that investment risk has been managed not by individual rational calculations[52] of risk and return but by following other market participants. Not only might this be ultimately damaging for the long term growth of corporations but it acts as a disincentive to investors to look at issues such as environmental management and environmental innovation within the corporation which are also long-term rather than short-term risks and to use their position as strategic influencers to change corporate behaviour and attitudes towards environmental degradation.

[48] See F Dobbin and D Zorn, 'Corporate Malfeasance and the Myth of Shareholder Value' (2005) 17 *Political Power and Social Theory* 179 and the pieces that follow this in the form of a debate: R Swedberg, 'On the Importance of Analyzing Economic Scandals and Contemporary Economic Institutions: A Comment on Dobbin and Zorn' ibid 199; E Clemens, 'The Power of Ideas? The Possibility of a Myth of Shareholder Value' ibid 207; and N Fligstein 'The End of (Shareholder Value) Ideology' ibid 223.

[49] 'A Review of Corporate Governance in UK Banks and Other Financial Industry Entities' (the Walker Report) November 2009, at 5.11.

[50] A Black and P Fraser, 'Stock Market Short-Termism—An International Perspective' (2002) 12 *Journal of Multinational Financial Management* 135.

[51] R Sias, 'Institutional Herding' (2004) 5 *Journal of Behavioral Finance* 181; and D Guyatt, 'Pension Collaboration: Strength in Numbers' (2008) 1 *Rotman International Journal of Pension Management* 46.

[52] R Parenteau, 'The Late 1990s', 'US Bubble: Financialization in the Extreme' in G Epstein (ed), *Financialization and the World Economy* (Cheltenham, Edward Elgar Publishing, 2005) 111, 126 provides a rich description of how investment practices lead to investment decisions that are very different from those that an efficient market hypothesis would predict.

Section 2

RESPONSIBLE OWNERSHIP

The phrase 'responsible ownership' or 'stewardship' has become particularly popular following the financial crisis of 2008, largely in respect of institutional investors and financial sector investee corporations, for example Barclays Bank and RBS. The general exhortations have been ones of using their voting power, rather than adopting a passive stance towards these rights, and engaging their investee corporations in dialogue. There are a number of Codes that have been put forward for institutional investors to follow in order that they may exercise their ownership function 'responsibly' by for example the Financial Reporting Council,[53] as the independent regulator, and professional representative associations such as the National Association of Pension Funds.[54] These have gradually been subsumed by a process of mutual recognition and support into the Stewardship Code. A full discussion of the Code[55] and its implications for institutional investors' intervention and collaboration[56] is not within the scope of this chapter not least because sadly it says nothing about the need for these investors to turn their attention to climate change and other issues of environmental or social concern despite its focus on action and engagement rather than passivity. The Code is designed to mirror the UK Code of Corporate Governance[57] and does so in that its seven principles cluster around dialogue,[58] disclosure of voting practices and monitoring, all of which are to be done in the pursuit of shareholder value.[59] Its main focus is on the firms that manage investment

[53] The UK Stewardship Code Financial Reporting Council, July 2010. This code incorporated the Institutional Shareholders Committee Code of November 2009 which itself built on the Myners Principles which followed the Myners Report (Institutional Investment in the United Kingdom: A Review) in 2001. As Myners pointed out in his 12 March 2009 conference speech to the National Pension Fund Association many of the problems of short-termism and lack of engagement that were addressed in his 2001 Report were still present in the stories of the financial crisis of 2008.

[54] Corporate Governance Policy and Voting Guidelines National Association of Pension Funds November 2007.

[55] For a fuller discussion see S Wong, 'Why Stewardship Is Proving Elusive for Institutional Investors' (2010) July/August *Butterworths Journal of International Banking and Financial Law* 406.

[56] I MacNeil, 'Activism and Collaboration Among Shareholders in UK Listed Companies' (2010) 5 *Capital Markets Law Journal* 419.

[57] M McKersie, 'The Stewardship Code and the Pattern of Engagement by Institutional Shareholders With Listed Companies' (2010) 5 *Capital Markets Law Journal* 439.

[58] J Logsdon and H Van Buren III, 'Beyond the Proxy Vote: Dialogues Between Shareholder Activists and Corporations' (2009) 87 *Journal of Business Ethics* 353.

[59] Shareholder value is mentioned only once in the code in Principle 4. Principle 4 states 'Institutional investors should establish clear guidelines on when and how they will escalate their activities as a method of protecting and enhancing shareholder value.' This exhortation can only really be made sense of if the protection of shareholder value is read into the other

assets for institutions. These firms are expected to disclose how they comply[60] with the Code through publishing information on their website. Institutions themselves are encouraged, as opposed to expected, to make available this information.[61] For the 40 per cent or so of the FTSE equity market that is foreign owned both the UK Code of Corporate Governance and the Stewardship Code represent aspirations only.[62]

FIDUCIARY DUTIES AND RESPONSIBLE OWNERSHIP

Responsible ownership and stewardship of investments framed in this way does not encourage anything by way of engagement with environmental concerns around the generation of profit by investee corporations. There has also been a view in common law countries that those occupying the role of fiduciaries (so primarily those individuals who are the trustees of pension funds) are prohibited from adopting an investment strategy for their trust/ fund that encompasses anything other than the 'beneficiaries .... best financial interests'.[63] This is drawn from a reading of the judgment of Megarry VC in *Cowan v Scargill*.[64] The facts of this case centred on a disputed investment strategy for the Mineworkers' Pension Fund. The trustees nominated by the National Union of Mineworkers would only accept an investment strategy that promised to end overseas investment and prohibit investment in any fuel which was in competition with coal. The coal industry in the UK was in its death throes at the time and the union mandated strategy appeared to be a political one of supporting the UK industry while failing to consider the effect that this investment strategy might have on the Fund's beneficiaries.

The key to reading the judgment would seem to be in recognising that what Megarry VC labelled a breach of fiduciary duty was the substitution of the beneficiaries' financial interests with the trustees' political beliefs. Environmental considerations were not mentioned in the judgment and no attempt was made by the National Union of Mineworkers to argue that its strategy could be justified for example in terms of protecting beneficiaries' long term financial interests. The best financial interests of the beneficiaries

principles which suggest actions that should be undertaken before the position referred to in principle 4 is reached.

[60] The parallel with the UK Code of Corporate Governance extends to the reference on page 1 of the Stewardship Code to 'comply or explain'. There is, however, no planned sanction for non-declaration of compliance or at least not in the way that there is in Corporate Governance Code. Presumably the market for asset manager's services will be left to decide the extent of the detriment incurred for non-disclosure

[61] This distinction is made clear in the preface to the Code.

[62] 'A Review of Corporate Governance in UK Banks and Other Financial Industry Entities' (the Walker Report) November 2009, at 5.41.

[63] [1985] 1 Ch 270 at 287.

[64] Above n 63.

as the sole objective of trustees' decision making has been questioned in a number of quarters,[65] not least by Megarry VC himself, writing in an extra-judicial capacity.[66] Further support for the reading of *Cowan v Scargill* given here[67] can be found in the change made to the regulations governing disclosure of investment policy by pension fund trustees in 2000.[68] This imposed a requirement that the trustees of occupational pension funds should include in their investment policy document a statement about the extent to which, if at all, social, environmental or ethical considerations were taken into account in the selection, retention and realisation of investments and a further statement about the way voting rights attached to these investment were exercised. There is a relatively well-supported view, legislation on policy disclosure notwithstanding, that institutional investors still feel inhibited/are inhibited from adopting an investment strategy that includes environmental concerns and engagement with other issues that raise ethical concerns.[69]

INSTITUTIONAL INVESTOR COALITIONS

However this is not to say that ideas of responsible ownership cannot be formulated which take account of environmental concerns and which capture the imagination of institutional investors even in an era where shareholder value has been aggressively promoted. The United Nations Principles

[65] See B Richardson, *Socially Responsible Investment Law* (Oxford, Oxford University Press, 2008) 228–34; and the United Nations Environment Programme Finance Initiative: A Legal Framework for the Integration of environmental, social and governance issues into Institutional Investment October 2005, commonly known as the Freshfields Report. Both of these sources provide very well argued support for the view that there is no judicial authority preventing environmental concerns from being part of an investment strategy or policy.

[66] R Megarry, 'Investing Pension Funds: The Mineworkers Case' in T Youdan (ed), *Equity, Fiduciaries and Trusts* (Toronto, Carswell, 1989) 149.

[67] See however R Thornton, 'Ethical Investments: a Case of Disjointed Thinking' (2008) 67 *Cambridge Law Journal* 396 where the diametrically opposing view is taken.

[68] Regulation 2(3)(b)(vi) of the Occupational Pension Schemes (Investment) Regulations 2005 (SI No. 3378/2005) issued under the Pensions Act 1995, section 35. Richardson, see above n 65, 304–09, offers a commentary on how disclosure regulations similar to this one have been enacted in numerous other jurisdictions: Australia, Germany, Italy and France for example.

[69] For example in November 2007 there was a dispute over investment policy between the councils of Liverpool and St Helens and the trustees of the Merseyside Pension Fund. The councils wanted the trustees to adopt a policy stopping the investment of funds in corporations that were involved in the arms trade. The pension fund trustees declined to adopt this policy on the ground that they were prevented from doing so by the judgment in *Cowan v Scargill* above n 50. www.responsible-investor.com/home/article/union_to_challenge_uk_law_on_ethical_investment/. The dispute was resolved away from the public gaze and the Merseyside Pension Fund website now declares that 'a strategy of engagement with companies, rather than screening companies from the portfolio on ESG grounds, is more compatible with its fiduciary duties', mpfmembers.org.uk/content/responsible-investment.

of Responsible Investment (UN PRI)[70] would be one such example. The UN PRI describes itself as an investor initiative in partnership with the United Nations Environment Program Finance Initiative and the United Nations Global Compact (itself an initiative aimed at the adoption of particular standards in relation to issues such as anti-corruption and human rights by the production function of corporations rather than the investment function). Launched in 2006 with principles developed by some twenty institutional investors, the UN PRI had some 550 signatories from both asset owners and investment managers representing around US$18 trillion in asset value in 2009.[71]

Unlike the Stewardship Code the UN PRI allows adoption of its principles with implementation reporting being conducted separately, in other words there can be adoption of the principles without any serious attempt on behalf of signatories to obtain disclosure from investee companies or to report on their own efforts at engagement. There are six principles for asset owners and their managers to sign up to, five of which are more concerned with investment process than any particular ethical stance around investment practises.[72] Only Principle 5 suggests taking a stance to achieve an outcome when it refers to a partnership approach to 'enhance our effectiveness in implementing the Principles'. There is no definition of responsible ownership; this is left to be divined as the factoring into investment decisions environmental, social and governance (ESG) concerns through the assertion in Principle 1 that these are things to be taken into account in investment analysis and decision-making processes.

It is clear from the supporting information to the Principles that ESG concerns are to be pursued to the extent that they deliver long-term financial value to beneficiaries.[73] This has the effect of avoiding any problems that might be perceived to exist in relation to the fiduciary duties of trustees and it ties ESG concerns firmly to business or material concerns.[74] ESG concerns are then rendered free of both values and political issues.[75] It is this tie that separates *responsible ownership* from what was once termed socially

[70] Principles for Responsible Investment, www.unpri.org/.

[71] *Fiduciary Responsibility: Legal and Practical Aspects of Integrating Environmental, Social and Governance* Issues Into Institutional Investment United Nations Environment Programme Finance Initiative otherwise known as *Fiduciary II*, 18.

[72] N Eccles, 'UN Principles for Responsible Investment Signatories and the Anti-Apartheid SRI Movement: A Thought Experiment' (2010) 95 *Journal of Business* 415.

[73] Above n 70, question 2 in Frequently Asked Questions section, www.unpri.org/faqs/.

[74] For a very interesting analysis of the journey from ethics to business in the environmental and social investment sector see B Richarsdon, 'Climate Financing and Its Governance: Moving to a Low Carbon Economy Through Socially Responsible Financing?' (2009) 58 *International and Comparative Law Quarterly* 597.

[75] M Welker and D Wood, 'Shareholder Activism and Alienation' (2011) 52 *Current Anthropology* S57, S65.

responsible investing[76] now better described as *ethical investing*.[77] There is no question of investee companies being subject to screening in relation to their products or production methods. This together with the incorporation of business concerns makes ESG investing palatable to a large number of institutional investors.[78] The extent to which institutional investors monitor their investee companies, seek out performance data from them and make suggestions for improvement is revealed only in UN PRI Progress Reports while allowing signatories to bask in the reputational capital acquired through association with a United Nations backed organisation.

The UN PRI is a distinctive organisation in that its focus is only on the institutional investor/investee company relationship, with asset managers included as part of the investment chain. There are other investor coalitions with broader membership bases—for example the Coalition for Environmentally Responsible Economies (CERES) with its Investors' Network on Climate Risk[79] and the Institutional Investors Group on Climate Change (IIGCC).[80] CERES is based in the US and has a membership drawn from institutional investors, NGOs and other environmental groups. It directs a group which calls itself the Investor Network on Climate Risk. This is group of more than 50 institutional investors from the United States and Europe.[81] IIGCC is based in Europe and from 2005 onwards it has been an independent part of an NGO—The Climate Group. Both of these groups are business interest NGOs (BINGOs) in that they go beyond the idea of responsible investment by institutional investors. They are not just fora for discussing principles of corporate governance and the repositories of information on disclosure by investee companies. They are BINGOs that lobby governments and policy makers for regulatory change and

[76] There are a number of institutional investors that have a policy of making investment decisions based upon a positive screening in or a negative screening out of particular corporations or industries based upon their end products or the production methods employed by them. For example the stated investment strategy might exclude corporations that make weapons, alcohol or tobacco products and include corporations which undertake to behave responsibly in relation to supply chain procurement. Others use a 'best in class' approach therefore permitting themselves to invest in industries which might otherwise be screened out. Some investment brokers have designed particular products for principally the individual retail market that target only screened funds or investee companies. Co-operative Financial Service figures for 2009 (constructed using information from the Investment Management Association and the British Bankers Association) place the amount of money held in ethical financial instruments at £19.2 billion up from £5.2 billion in 1999. Although showing strong growth, this figure represents only 1.8% of amounts held in UK based financial instruments.

[77] S Viviers and J Bosch, 'Is Responsible Investing Ethical?' (2008) 39 *South African Journal of Business Management* 15.

[78] C Woods and R Urwin, 'Putting Sustainable Investing Into Practice: A Governance Framework for Pension Funds' (2010) 92 *Journal of Business Ethics* 1, 4.

[79] www.ceres.org; www.incr.com.

[80] www.iigcc.org.

[81] J Galbreath, 'Corporate Governance Practices That Address Climate Change: An Exploratory Study' (2010) 19 *Business Strategy and the Environment* 335.

enforcement on environmental issues. Their focus is the risk to their equity holdings that is represented by regulatory change.[82] They are interested in creating a space in which the risks posed by this can be effectively managed as part of a risk management strategy by investee companies.[83]

The reasons for the emergence of institutional investors' interest in ESG issues tied to financial performance are beyond the scope of this chapter simply by means of the length of the explanation that would be required. In short form, however, this interest should not be seen as any great conversion to the cause of environmental or other social issues but more as a support to the protection and enhancement of the brand value of their investee companies.[84] Just as the late 1980s and early 1990s saw the realisation that a market advantage could be created from the adoption of innovative and relatively low cost corporate social responsibility policies[85] so the first decade of this century has seen the same realisation dawn in relation to environmental concerns. Corporate responses to state and supra-state level regulation on climate change are seen as creating both risks and opportunities for investee companies.[86] Institutional investor monitoring of investee companies responses' to regulation allows an informed investment strategy to flourish and potential risks to be identified and managed.[87] Lobbying for effective enforcement is about ensuring that corporate innovation is rewarded and perhaps more importantly squeezing out competitors for whom the compliance costs are too high. Intervention in the policy-making process allows regulation to be shaped by corporate interests.

INDIVIDUAL SHAREHOLDER ACTIVISM

The term shareholder is used here to represent either the actions of institutional investors when they are acting as individual actors and not through the representative offices of a coalition or the activities of individual asset

[82] See, for example, the alliance that was formed in 2008 between US listed corporations—Nike, Starbucks, Levi Strauss, Sun Microsystems and Timberland. Called Business for Innovative Climate and Energy Policy (BICEP) their launch document called for strong US climate and energy legislation to be adopted. BICEP is now a project of CERES and has attracted additional members www.ceres.org/bicep.

[83] J Hawley and A Williams, 'Shifting Ground: Emerging Global Corporate-Governance Standards and the Rise of Fiduciary Capitalism' (2005) 37 *Environment and Planning* 1995.

[84] G Clark and T Hebb, 'Why Should They Care? The Role of Institutional Investors in the Market for Corporate Global Responsibility' (2005) 37 *Environment and Planning A* 2015, 2024.

[85] C Smith, 'The New Corporate Philanthropy' (1994) 72 *Harvard Business Review* 105.

[86] J Llewellyn and C Chaix, 'The Business of Climate Change II' (Lehman Bros, 2007).

[87] *Investor Progress on Climate Risk and Opportunities* (CERES, 2009); *Climate Changes Your Business* (KPMG, 2008); M Lubber, 'Risks and Their Impact on Institutional Investors' in A Calvello (ed), *Environmental Alpha: Institutional Investors and Climate Change* (Hoboken, NJ, Wiley, 2010) 79.

management firms which have formulated their own operating principles for investment. Hermes Fund managers which act on behalf of the BT pension fund, for example, have created *The Hermes Principles*.[88] These are very similar to the UK Code of Corporate Governance and the Stewardship Code. Environmental, ethical and social issues are the subject of Principles 9 of 12. These issues are to be considered in the context of relationships with 'employees, suppliers and customers and with others who have a legitimate interest in the company's activities' while ensuring that value creation remains the operating norm. Interestingly Principle 9 goes on to warn of the need for a robust stance by investee companies when dealing with 'sectional interests'. Individual codes of this type aside, activism here refers to 'visible intervention' in an investee corporation's strategy through either putting forward a shareholder resolution which is at odds with the position taken by the particular corporation's management or through voting against a resolution put forward by managers: resolution based activism. NGOs such as Greenpeace have in the past purchased shares in a corporation so that they can either propose a resolution or vote against a particular resolution. Standing outside this definition of activism is any dialogue entered into as a result of the Codes referred to above that falls short of visible intervention in the context of shareholders' meetings.

All national regulatory regimes for corporations have mechanisms by which shareholders can call a shareholders' meeting for the purposes of passing a resolution or by which shareholders can place a resolution on the agenda of the annual general meeting of shareholders. These mechanisms differ across national regimes thus making the measuring of levels of shareholder participation in resolution-based activism difficult. The US, where the most detailed data on resolution filing is available, has a shareholder access regime that is quite limited in comparison with, for example, the UK or Australia[89] in that the corporation concerned can request that the SEC approve omission of the resolution on one of the 13 grounds available to it. These range from the subject matter of resolution being considered as not significantly related to the corporation's business to a conclusion that the company has already substantially implemented the proposal.[90] In the US in 2008 there were 64 climate change shareholder resolutions filed,[91] a steady increase from less than 10 in 2000.[92]

[88] www.hermes.co.uk/Portals/0/The_Hermes_Ownership_Principles.pdf.

[89] J Hill, 'The Rising Tension Between Shareholder and Director Power in the Common Law World' (2010) 18 *Corporate Governance* 344; Securities Exchange Act (1934) Rule 14a-8.

[90] E McAteer and S Pulver, 'The Corporate Boomerang: Shareholder Transnational Advocacy Networks Targeting Oil Companies in the Ecuadorian Amazon' (2009) 9 *Global Environmental Politics* 1.

[91] E McAteer *et al Sustainability Background Report: Global Climate Change* (RiskMetrics Group, 2009).

[92] It should not be forgotten that the use of resolution-based activism has a long and honourable history on social issues in particular. The campaigns around apartheid in SA and

6

Climate Change, Environmental (In)Security, Conflict and Crime

NIGEL SOUTH

INTRODUCTION

WE LIVE IN a world brought closer together yet at the same time divided and brought into conflict through the processes of globalisation. In this late-modern condition, not only must we come to terms with and be able to respond to the crises caused by natural disasters beyond human control[1] but we must also find better ways to reduce the damage caused by our own actions as exploiters of nature, manufacturers of waste and as intemperate consumers. In this respect, the problems we face are also part of what Beck[2] called 'the risk society', whereby modern industrial societies create new risks (largely manufactured via the advance of technologies) that were unknown in earlier decades. Commenting on the ways in which the dominant discourses of risk and threat have changed over time, Dunne[3] observes that:

> Every age seems to invent a master narrative about its own demise. At the turn of the 20th century, the fear was of a violent struggle between races and civilisations. In the post-1945 world, the nightmare was of a nuclear holocaust. The narrative of demise that haunts us today is the threat that environmental harm holds for the planet and its ability to support human and other life forms.

Perhaps more precisely, it is the prospect of the intensification of climate change that has become the basis for the global 'narrative of demise'. The consequences of this can be perceived and anticipated in various ways. For example, climate change will shift the patterns and transactions of both

[1] P Blaikie, T Cannon, I Davis and B Wisner, *At Risk: Natural Hazards, People's Vulnerability and Disasters* (London, Routledge, 2003).

[2] U Beck, *The Risk Society* (London, Sage Publications, 1992).

[3] T Dunne, 'How Do We Secure the Environment: Review of "Security and Environmental Change" by S Dalby, Polity, 2009' *Times Higher Education Supplement*, 8 October 2009, 46.

local and global life, both encouraging and impeding mobility, as well as leading to the revaluation of resources and the restructuring of markets, and as one outcome (among many) also re-shaping the causes and nature of crime and conflict.

A growing body of work on (variously) 'green', 'eco' or 'environmental' criminology is now engaged with the issues and problems posed by harms to the planet and the species that share it, making interdisciplinary connections with other investigators tackling similar topics from differing intellectual and methodological directions, and providing a framework in which an agenda for future research can evolve.

THE DEVELOPMENT OF A 'GREEN PERSPECTIVE' IN CRIMINOLOGY

There is a long record in criminology and socio-legal studies of writers exposing failures of regulation and avoidance of responsibility regarding the preservation of the environment.[4] However, over the past twenty years, paralleling a general increase in awareness regarding threats to nature and human health caused by the over-exploitation of resources and the impacts of pollution, a body of work explicitly informed by and concerned with green issues has emerged.[5] Originally proposed as a 'green criminology'[6,7] the appropriate name or label for this sub-field or perspective has subsequently been subject to some debate so, for example, White[8] now prefers the term environmental criminology as appropriate for the study of environmental harms and threats, environmental legislation and related research activity.[9] Walters[10] has suggested that the term 'eco-crime' is useful and capable of encapsulating 'existing legal definitions of environmental crime, as well as sociological analyses of those environmental harms not necessarily specified by law'. Conceived in this way, Walters[11] suggests that we can accommodate discourses on risk, rights and regulation.

[4] N South and P Beirne (eds), *Green Criminology* (Aldershot, Dartmouth Publishing, 2006).

[5] P Beirne and N South (eds), *Issues in Green Criminology: Confronting Harms Against Environments, Humanity and Other Animals* (Cullompton, Willan Publishing, 2007).

[6] M Lynch, 'The Greening of Criminology: A Perspective on the 1990s' (1990) 2 *Critical Criminologist* 3–4, 11–12.

[7] N South, 'A Green Field for Criminology?: A Proposal for a Perspective' (1998) 2 *Theoretical Criminology* 211–34.

[8] R White, *Crimes Against Nature: Environmental Criminology and Ecological Justice* (Cullompton, Willan Publishing, 2008) 8.

[9] R White (ed), *Global Environmental Harm: Criminological Perspectives* (Cullompton, Willan Publishing, 2010).

[10] R Walters, 'Eco Crime' in J Muncie, D Talbot, and R Walters (eds), *Crime: Local and Global* (Devon, Cullompton, 2010) 180.

[11] Ibid, 181.

In fact all these terms represent similar frameworks and share a common interest in the bio-physical and socio-economic consequences of various sources of threat and damage to the environment whether pollution, resource degradation, biodiversity loss or climate change.[12] Furthermore, whatever the descriptor, the subject matter is fairly well agreed upon: for example, pollution and its regulation; corporate criminality and impacts on the environment; health and safety in the workplace where breaches will have environmentally damaging consequences; involvement of organised crime and corrupt officials in the illegal disposal of toxic waste; the influence of law enforcement and military operations on landscapes, water supply, air quality and human populations; and forms of law enforcement and rule regulation relevant to such acts.[13,14]

Climate Change, Crime and Conflict

In what follows, I want to focus on climate change as the basis for the kind of global 'narrative of demise' referred to earlier and, in particular, to consider how (contingent but plausible) developments might then give rise to new formations of global flows of people and goods,[15] probably shaped by new forces of order and exclusion. In this scenario, environmental *in*security defines the political and policy mood, with implications for international and national security regimes, as well as for crime and conflict. Let me start by working through some of the changes, local and global, that we might anticipate and that will therefore influence this mood.

Water, Food and Scarcity

As Revkin[16] reports

> Scientists say it has become increasingly clear that worldwide precipitation is shifting away from the equator and toward the poles. That will nourish crops in warming regions like Canada and Siberia while parching countries—like Malawi in sub-Saharan Africa—which are already prone to drought.

[12] M Hauck, Non-Compliance in Small-Scale Fisheries: A Threat to Security?' in P Beirne and N South (eds), *Issues in Green Criminology* (Cullompton, Willan Publishing, 2007) 272.

[13] South, above n 7, 214.

[14] K Eman, G Meško and C Fields, 'Crimes Against the Environment: Green Criminology and Research Challenges in Slovenia' (2009) 11 *Journal of Criminal Justice and Security* 574–92.

[15] KF Aas, *Globalization and Crime* (London, Sage Publications, 2007).

[16] A Revkin, 'Poor Nations to Bear Brunt as World Warms' *New York Times*, 1 April 2007.

> While rich countries are hardly immune from drought and flooding, their wealth will largely insulate them from harm, at least for the next generation or two, ...

Controlling water means controlling life. There is ample scope here for political corruption, corporate irresponsibility, even genocide. Clean water is essential not only for drinking and preparing food but also as an elementary but effective means of helping to control the effects of pollution and disease. However, if either of the latter dangers should compromise the purity and value of water then conditions can deteriorate in ways that threaten stability and security. If the risks of such contamination and devaluation increase, whether due to manufactured or natural causes, then the market price for alternative supply will increase, to be met via legal or illegal, peaceful or hostile, means. A new world of hydropolitics emerges in situations where water sources are currently accessed by several nations but could potentially be controlled—or indeed monopolised—by one nation or by private water and power consortia. Such conditions already produce tensions but in addition, as French argues:

> There is no doubt ... that future conflicts will centre upon states' need—real and perceived—for access to freshwater resources.[17]

Climate change will produce scarcity which means rationing, in turn producing conditions that support corrupt diversion, illicit markets, profiteering, food riots and human security challenges for policing and distribution agencies. Drought-induced food scarcity obviously has consequences of a devastating nature in terms of human casualties and human rights but is also criminogenic as such conditions can give rise to illicit markets in food and medicines. At the other extreme of environmental conditions, flooding destroys food crops, pollutes and spreads disease and similarly gives rise to the need for aid and the search for ways of escape which, in turn, offer criminal market opportunities.

Environmental rights, human rights and 'environmental security' come together around the matter of access to water and food as fundamental requirements for life. Ultimately, climate change will produce differentiated social vulnerability to water and food scarcity[18] and one pattern of response may be increased migration.

Borders, Migration and Belonging

Borders are generally permeable to trade even if more restrictive of human population movements. In 2010, food protests in Mozambique provided

[17] D French, 'Environmental Security in an Insecure World' (2005) 17 *Environmental Law and Management* 159.

[18] H Bohle, T Downing and M Watts, 'Climate Change and Social Vulnerability: Toward a Sociology and Geography of Food Insecurity' (1994) 4 *Global Environmental Change* 37.

an illustration of how the interaction of changing weather patterns in different agricultural regions and influences on the commodity markets that transcend borders can cause conflict and crime. The riots and robbery of grocery stores followed a period of price increases for bread as well as for water and electricity supply. In turn, the rising bread prices reflected the inability of local crop production to meet demand which could no longer be easily matched by purchase from abroad when wildfires destroyed parts of the grain belt of Russia during the hottest summer heat-wave for a century, leading to export bans from that country. At the same time, international commodity speculators were engaging in what the World Development Movement has called 'gambling on hunger in financial markets'.[19] Such localised crises could tip over into full-scale regional disasters with the addition of only a few further destructive pushes such as, in some areas, rising sea levels introducing salt water onto agricultural land at the same time as fresh water shortages impact on farming. The resulting food crises and starvation would be likely to drive many refugees onto the road of migration. In terms of destinations, Europe and the UK in particular are already key objectives and likely to become more so.[20]

Not only is migration increasingly likely to be driven by inter-related conditions of poverty, human conflict and environmental crises but in such circumstances all that is entailed can also lead to further damage and despair. The process of losing a home and sense of place, being vulnerable and without shelter or focus while undertaking a journey of uncertainty, can lead to victimisation and crime. In particular, the psycho-social impact of displacement and sense of 'loss' makes people vulnerable to exploitation—whether encouragement into crime, capture into human trafficking for sex and labour, or recruitment as disposable soldiers in wars that are often waged over the riches of the earth, in turn displacing more people. Among the key issues that follow is the degree to which more advantaged nations keep faith with the sense of 'humanitarian responsibility' that might normally be expected in times of disaster and difficulty. There is evidence of both enormous charitable funding and practical assistance being extended in such circumstances, from rich to less wealthy nations. However, 'humanitarianism' is politically sensitive. Given that there is also substantial evidence about how easily 'compassion fatigue' can emerge in these same circumstances it would not be a surprise to find high levels of resistance to perceived over-extension of assistance to increased numbers of migrants from developing to developed nations. McKie argues that:

> Britain is likely to be one of the few nations to survive the worst effects of climate change while other nations, particularly those in the developing world, have their

[19] R Patel 'Mozambique's Food Riots—the True Face of Global Warming' *The Observer*, 5 September 2010, 31.

[20] R McKie, 'Britain Set to Become Most Populous Country in EU' *The Observer*, 22 March 2009.

> farmland and fishing grounds destroyed. It could be argued that the UK has a moral duty to provide shelter for as many refugees as our shores can support. But deciding what numbers the country might support is a highly controversial issue.

Given that conflicts related to ethnic and cultural difference have been recurrent challenges to urban life across Europe for many years, it is reasonable to predict that increased inward migration pushed by climate change could be a source of strain that will exacerbate existing tensions in some communities. In this case, ideals of cosmopolitanism and tolerance may fall out of line with the preoccupations of nationalist politics and the priorities of national security, turning those who should be seen as 'deserving victims' into 'transgressors' and 'threats'. Such circumstances may have the force required to reverse any momentum toward acceptance and implementation of human rights via trans-national developments associated with the expression of cosmopolitanism. In this regard, Morris[21] argues that, 'With the incorporation of international legal norms into domestic regimes, we have a concrete case of the blurring of internal and external factors, seen by some as the basis for a global citizenry' and hence a world that invites and supports mobility across borders and boundaries. However, if human rights become devalued and subjugated to other imperatives—nationalism and survivalism—then borders will become less permeable and boundaries regarding tolerance and diversity could become tighter. In addition, the question of who is deserving of that which has become scarce will be answered by measures of 'belonging'. From here we are not far from thoughts that support a principle of preserving a specific place for a specific population. Migration pushed by climate change elsewhere might be expected to generate some sympathy and as Morris[22] suggests, in some cases it can be argued that people may feel that by 'defending the human rights of foreigners and strangers ... they are defending their own identities'. Yet, as Morris[23] also notes, Douzinas[24] has registered the way in which 'the plight of the refugee puts claims of the universalization of rights to the test' and that Habermas[25] refers to the 'peculiar tension' that can arise 'between the universal meaning of human rights and the local conditions of their realization'. Climate change-induced migration on a large scale will amplify such tension in equally large proportion. The consequences of such hyper-nationalism and tough protectionism will be important politically, socially and criminologically.

[21] L Morris, 'An Emergent Cosmopolitan Paradigm? Asylum, Welfare and Human Rights' (2009) 60 *British Journal of Sociology* 216.

[22] Ibid.

[23] Morris, above n 22.

[24] C Douzinas, *The End of Human Rights* (Oxford, Hart, 2000).

[25] J Habermas, *The Post-National Constellation* (Cambridge, Polity Press, 2001).

Environmental (In)Security, Prevention and Response

There are some key questions that follow from the preceding observations. Two of these are how well prepared can we be for the consequences of climate change in general and those generating crime and conflict impacts specifically? Are there investments that can be made of a preventive nature and/or to improve capacity and readiness to respond? A programme of research that would aim to address such questions is an obvious response and an inter-disciplinary approach that fostered complementary inquiry in other fields should be encouraged. But in addition, there will be substantial scope to learn lessons from studies of prevention and responses in the case of natural disasters and other challenges posing similar problems. Take, for example, recent work on the optimisation of efforts to reduce mortality in the event of earthquakes.

Briefly, if we take earthquakes as a naturally occurring phenomenon, hence different from human-induced climate change but accompanied by similar challenges to socio-economic systems and threats to populations (mortality, casualties, migration), what does the interdisciplinary study of earthquake mortality prevention carried out by Keefer *et al*[26] tell us? According to these researchers, there is a divide between what rich and poor nations can and will find it most effective to do. The idea of such a divide is familiar, the actual responses perhaps less so. The response of a government to the likelihood of earthquakes 'depends on country income and the political incentives of governments to provide public goods to citizens.' Interestingly (though perhaps not surprisingly) nations with less democratic and open systems of governance—institutionalised autocracies and corrupt regimes—*will invest less even where there is an elevated level of earthquake likelihood.* The criminological implication is that—whether faced by earthquake or climate change—deviant or criminal states will be less compliant with normative or legislative expectations about investment and action required for prevention or preparation for response. This obviously reflects internal stances toward climate change and the conditions that cause environmental damage but it also has implications for the view that 'development' is the key to turning such states of insularity and inaction into parties to collective (internal and international) response. For example, French[27] takes an extract from the 2004 UN High Level Panel report on Threats, Challenges and Change that has just such an emphasis, whereby 'development' is seen as 'the indispensable foundation for a collective security system that takes prevention seriously.' There are 'multiple

[26] P Keefer, E Neumayer and T Plumper, *Earthquake Propensity and the Politics of Mortality Prevention* (2009), Social Science Research Network Working Paper Series.

[27] French, above n 17, 160.

functions' served by the development process and these can be summarised as follows:

> It helps combat the poverty, infectious disease and environmental degradation that kill millions and threaten human security. It is vital in helping States prevent or reverse the erosion of State capacity, which is crucial for meeting almost every class of threat. And it is part of a long-term strategy for preventing civil war, and for addressing the environments in which both terrorism and organized crime flourish (part ii, synopsis).

However, development as a prevention measure requires investment and while it may stem civil war and possibly weaken gangsterist regimes it will also be impeded by both legitimate and illegitimate factors. In the case of criminal states that despoil and exploit their natural resources and care little about global warming, the interests of those in power are best secured by inhibiting development, except in specified and controlled ways. Having noted this, it is salutary to also observe that even highly democratised societies, those of the west with free markets and de-regulated banking and finance systems, are not always hugely credible when it comes to their own track records on environmental investment. And, to join the developed and developing worlds together, this is particularly so in the case of investment by some western multi-nationals in developing nations where, as Newell notes,

> the mobility of capital and the internationalization of production that make international investment possible, give companies unprecedented freedoms to locate their businesses where it is most profitable to do so, often at the expense of communities and their environment ... Of particular concern is that developing countries often experience greater economic and political volatility which means that foreign investors tend to engage in ventures that will yield a high rate of return over a short period, often resulting in environmental devastation and social dislocation.

Such conditions will contribute to the creation and impact of 'environmental insecurity'.[28]

EXPLORING THE CONCEPT OF ENVIRONMENTAL INSECURITY

Evidently environmental change and harms pose a challenge to the security and sustainability of nations and their populations but the form and style of appropriate response is not easily settled upon.[29] Typically in the past, governments have responded to threats to security (usually national) and

[28] P Newell, 'Managing Multinationals: the Governance of Investment For the Environment' (2001) 13 *Journal of International Development* 907.

[29] F Dodds and T Pippard, *Human and Environmental Security: an Agenda for Change* (London, Earthscan, 2005).

sustainability (usually of supplies) through the marshalling of armed forces and deployment of technologies of conquest and war. In terms of current sources of global anxieties such a response may be felt by some (though not all) to be inappropriate and this was recognised over 30 years ago in the Brandt Report[30] when the authors noted that

> True security cannot be achieved by a mounting build up of weapons (defence in a narrow sense), but only by providing basic conditions for solving non-military problems which threaten [human life]. Our survival depends not only on military balance but on global cooperation to ensure a sustainable environment.[31]

Indeed, in the way that it could be argued that a truly successful national security policy should result in disarmament, so by extension the aim of a successful environmental security policy should be mitigation of the problem of climate change.[32]

To achieve such mitigation would require collective action across borders, in effect, an effort to promote collective and mutual environmental security. To move in this direction, a sustainable environment needs the support of a different version of 'security'—a 'sustainable security' as Voigt[33] puts it:

> based on the understanding that security cannot only be political or military, but that it must at the same time be social, economical and ecological. Neither can it be purely national; it needs to be collective and inclusive. Only then can security evolve to be universal.

Of course, the concept of 'security' whether public, private or hybrid, is by no means new in criminology and has been explored in various ways by, for example, Zedner,[34] Loader and Walker,[35] South,[36] Johnston and Shearing.[37] Yet because the study of environmental harms and green issues *is* fairly new to criminology, the idea of '*environmental* security' has hardly been discussed at all in the criminological literature. Where it is found with reference to associated crime, whether within or outside criminology texts and journals, the authors themselves tend to be external to the field.[38,39]

[30] The Brandt Report, *North–South: A Programme for Survival* (Cambridge, MA, MIT Press, 1980) 119.

[31] See also the 1992 Rio Declaration quoted in French above n 17, 159.

[32] I am grateful to Catherine Redgwell for making this point at the Oñati seminar.

[33] C Voigt, 'Sustainable Security' in OK Fauchald, D Hunter and W Xi *et al* (eds), *Yearbook of International Environmental Law* 19 (Oxford, Oxford University Press, 2009) 164–65.

[34] L Zedner, *Security* (London, Routledge, 2009).

[35] I Loader and N Walker, *Civilizing Security* (Cambridge, Cambridge University Press, 2007).

[36] N South, *Policing for Profit: the Private Security Sector* (London, Sage, 1988).

[37] L Johnston and C Shearing, *Governing Security: Explorations in Policing and Justice* (London, Routledge, 2003).

[38] M Hauck, 'Non-Compliance in Small-Scale Fisheries: A Threat to Security?' in P Beirne and N South (eds), *Issues in Green Criminology* (Cullompton, Willan Publishing, 2007).

[39] L Elliott, 'Transnational Environmental Crime in the Asia Pacific: an "Un(der)securitized" Problem' (2007) 20 *Pacific Review* 499–522.

The concept has received somewhat greater attention in law[40,41] but there is still much to explore in the territory where matters of environmental harms, crime and conflict all meet and overlap with questions concerning security, rights and law.

Little and Cocklin[42] have remarked that while consensus about terms such as 'sustainable development' has remained somewhat elusive, there is now widespread and agreed recognition of the interdependence between humanity and the natural environment and that, while less well known, the recognition of a connection or nexus between environment and security is complementary to this. French[43] notes that the 2004 report of the UN High Level Panel on Threats, Challenges and Change suggests that '[r]arely are environmental concerns factored into security, development or humanitarian strategies' (paragraph 54). Yet while the development of the relevant 'strategies' may have been slow to develop, the concept and discourse of environmental security is not entirely new as Hauck[44] records, becoming established in the mid-1980s with renewal of interest in the mid-1990s[45] to the point where perhaps it is now a more routinely inserted element of debates and strategies.

In a successful example of an inter-disciplinary approach to discussion of such threats and challenges, Hauck[46] applies the concept of security to a case study of the conflict and crime associated with fisheries management during the changes occurring in post-apartheid South Africa. Hauck's essay is a contribution to a collection on issues in green criminology and it is therefore appropriate that she draws attention to the 'little direct reference' [in the green criminology literature] 'thus far to the discourse on human security, non-traditional security, or more specifically, environmental security'. Arguing for the need to remedy this, Hauck describes the broadening of the concept of 'security', from a traditional association with the ability to counter external military threats to the embrace of broader political, economic, social and environmental concerns, and recognition not just of the need for protection of the state but also of individuals and their communities.[47,48] The concept of 'human security' Hauck notes, was given

[40] French (n 17).

[41] K Hulme, 'Environmental Security: Implications for International Law' in OK Fauchald, D Hunter and W Xi (eds), *Yearbook of International Environmental Law* (Oxford, Oxford University Press, 2008).

[42] L Little and C Cocklin, 'Environment and Human Security' in J Pretty *et al* (eds), *The Sage Handbook of Environment and Society* (London, Sage Publications, 2007) 442.

[43] French (n 17) 159.

[44] Hauck (n 38) 272.

[45] N Myers, 'Environmental security: what's new and different?' Background paper for The Hague Conference on Environment, Security and Sustainable Development (2004).

[46] Hauck (n 38) 270.

[47] Hauck (n 38) 271.

[48] L Little and C Cocklin, 'Environment and Human Security' (2007) 442–44.

emphasis by the 1994 Human Development Report of the United Nations Development Programme (UNDP, 1994) with the subsequent Commission on Human Security (2004) emphasising 'that human security is about protecting people's fundamental rights, such as freedom, peace and safety, access to resources and the basic necessities of life' and—importantly—that this also encompassed 'an environment that does not threaten health and well-being'.

There are however, other ways of approaching the nexus of human and environmental security and these can take a bleaker turn. Writing first in a highly influential essay published in 1994, and then in a subsequent book in 2000 that amplified the themes, the American political commentator Robert Kaplan[49] described challenges and changes associated with the global environment as the source of 'The Coming Anarchy'. This is a far more parochial, human- and American-centred focus, with Kaplan addressing an agenda where the main preoccupation is the 'national security' of the United States. Kaplan's comments remain worthy of note. Here in one far-sighted section of his article, titled 'The Environment as a Hostile Power', Kaplan argued that

> It is time to understand 'the environment' for what it is: *the* national-security issue of the early twenty-first century. The political and strategic impact of surging populations, spreading disease, deforestation and soil erosion, water depletion, air pollution, and, possibly rising sea levels in critical, overcrowded regions like the Nile Delta and Bangladesh—developments that will prompt mass migrations and, in turn, incite group conflicts—will be the core foreign-policy challenge from which most others will ultimately emanate ...

Whether environmental matters really are moving toward the true core of foreign policy concerns is a point open to debate but Kaplan's argument, as well as some of the critical responses in the literature, are worth dwelling on. Here it is worth noting the discussion of Kaplan's stance by Dalby[50] who described it as a 'rhetorically powerful analysis' and a:

> a high-profile public articulation of contemporary neo-Malthusian themes in post-cold war geopolitical discourse. It parallels much of the rest of the U.S. media coverage in its representations of Africa, and Rwanda in particular, as a place of tribal, hostile, violent Others. It is notable for its pessimism, forceful prose, and the absence of any suggested substantive political remedies to the immanent dystopia.

Kaplan found support from quarters exhibiting deep anxieties about crime, the aftermath of the cold war and the environment, a mixed political bag including both Republicans and the White House administration of the Clinton era. Taken as clear and powerful journalism (Kaplan had been a

[49] R Kaplan, *The Coming Anarchy* (New York, Vintage Books, 2000).
[50] S Dalby, *Environmental Security* (Minneapolis, University of Minnesota Press, 2002) 22.

war correspondent and travel writer) Kaplan's work has to be admired for its effectiveness. His approach is informed by the logic of Malthusian arguments and evidence of famines in Africa and elsewhere during the 1980s and struck a chord with similarly-inclined conservative commentators but as Dalby[51] notes, 'the powerful rhetoric of the argument retraces familiar political territory instead of looking in detail at the environment as a factor in social change.'

The underlying message here is not about what humanity should do to prevent hostile and irreversible actions against the environment but about how the environment itself is becoming a hostile force that will work against the national security and foreign policy interests of the USA. Dalby[52] observes that Kaplan sees the hostility of the environment in terms of emergent 'wild zones' that form part of the geo-political dis-order of contemporary global formations and that have the potential to generate 'disruptive incursions into the "tame" zones of postmodern prosperity', requiring containment—'if necessary by military force'. Kaplan clearly operates with a US-centric dualism that splits the world into civilised and primitive, functional and dysfunctional. Urging avoidance of the twin dangers of contamination (of the west via over-exposure to the problems of Africa) and of failure (to intervene in any successful way on this continent) leads Kaplan to argue for separation between the worlds of developed and developing nations and the need for barriers that will prevent the transfer of troubles from Africa to elsewhere. As Dalby[53] puts it, 'once again security is understood in the geopolitical terms of containment and exclusion'. This is a model that underpinned the colonialism that empowered earlier waves of environmental—and human—exploitation, the removal of resources from one continental location to another and justified past measures designed to regulate or prevent migration and movement. This was never a defensible approach to conceptualising and operationalising security and cannot be conceived as such today.

What is vital in contemporary calculations of sustainability and security is the factor of change. Security, as Dalby[54] argues, 'is not about protecting a stable status quo from an external threat. It is about developing an economic system that reduces dependence on a single resource, a dynamic system that can accommodate change. This does not fit easily into traditional understandings of defence or national security, which is, of course, the whole point of trying to rethink security.' One way to do this 'rethinking' has been to

[51] Ibid, 36.
[52] Ibid, 26.
[53] Ibid, 38.
[54] Ibid, 60.

explore the idea of environmental security in the context of the pre-eminent change of our times. As the climatologist, Hulme[55] writes

> … climate change has come to signify far more than the physical ramifications of human disturbance to the composition of the Earth's atmosphere and its energy balance. … Arguments about the causes and consequences of climate change—and the solutions to it—have become nothing less than arguments about some of the most intractable social, ethical and political disputes of our era: the endurance of chronic poverty in a world of riches; the nature of the social contract between state and citizen; the cultural authority of scientific knowledge; and the role of technology in delivering social goods. Climate change has become a metaphor for the imagined future of human life and civilisation on Earth.

Climate change is producing a new set of global dividing lines, now between those at most risk and those at least risk. This 'climate divide' is recognised in many ways but arguably not on a widespread basis or with full appreciation of what it really means. In essence, the climate divide represents a further extension of the inequitable state of the affairs of humanity, one in which the conditions producing climate change are contributed to most overwhelmingly by rich consumer societies but which will impose the greatest costs and resultant miseries on the already poor and newly developing nations.

In a feature in the *New York Times* on the effect of global warming on the poorest nations, Revkin[56] quotes Rajendra K Pachauri, chairman of the United Nations Climate Panel as saying 'The inequity of this whole situation is really enormous if you look at who's responsible and who's suffering as a result'. In response, says Revkin, 'Those in harm's way are beginning to speak out: "We have a message here to tell these countries, that you are causing aggression to us by causing global warming"' President Yoweri Museveni of Uganda said at the African Union summit in Addis Ababa, Ethiopia, in February. This interpretation of the causes (and causers) of climate change as manifestations of aggressive action is obviously relevant to attempts to employ the notions of 'security' and 'environmental security' in sensitive ways that move the discourse away from threat and hostility.

CONCLUSION

It might be argued that the concept of environmental security is simply stretching the 'security' umbrella too far because it encompasses (and blurs with) many other and well-understood aspects of security—political, economic and social. Inevitably, it is also entwined with social injustice,

[55] M Hulme, 'Heated Debate' (2010) 5541 *RSA Journal* 36.

[56] A Revkin, 'Poor Nations to Bear Brunt as World Warms' *New York Times*, 1 April 2007.

poverty, differential vulnerability, weak political structures, population growth, unsustainable economies, industrialisation and resource demand. However, it is absolutely right that conceptual understanding should be so stretched. There are certain obvious fundamentals that need to be secure for the future of the human race, other species and the planet. The approach to security priorities must reflect this and incorporate the capacity to nurture and develop in sustainable ways, rather than re-emphasising the forms of security underpinned by threat, violence and potential annihilation. This means we need fresh thinking about what security means at the inter-state level but it may also be fruitful to try to re-interpret the idea of security using ideas more often applied at the inter-personal level. For example, this might be done in ways more familiar in the fields of psycho-analysis and the sociology and psychology of childhood or health rather than always retaining the default usage common in the fields of political science and international relations. It is, after all, the very essence of human security that is shattered when faced by the trauma of conflict or the shock and loss that follow disaster—'when human beings are confronted with the limits of their existence as subjects'.[57]

An interdisciplinary perspective is needed to help move toward the necessary policy and practice here and collaboration across research boundaries will also play a role. One illustration, a report on 'Managing the Health Effects of Climate Change',[58] has its origins in an international public health study but can easily be seen to connect with the concerns of law, philosophy and criminology relevant to the challenges of climate change:

> Management of the health effects of climate change will require inputs from all sectors of government and civil society, collaboration between many academic disciplines, and new ways of international cooperation that have hitherto eluded us (p1693 col 2). ... Climate change also raises the issue of intergenerational justice. The inequity of climate change—with the rich causing most of the problem and the poor initially suffering most of the consequences—will prove to be a source of historical shame to our generation if nothing is done to address it (p1694 col 1).

Criminology, law and public health share a common interest in arguing for the protection of the individual and of the rights that form the links in the chain of intergenerational justice. Some have proposed that the notion of inter-generational justice or equity between generations be translated into identifiable obligations and transmission of rights that would be underpinned

[57] J Edkins, 'Security, Subjectivity and Trauma: Psychoanalytical Concepts in International Security', ESRC Final Report, Award No R000222190 (1999) 3.

[58] A Costello, *et al*, 'Managing the Health Effects of Climate Change' (2009) 373 *The Lancet* 1693–94.

by international law. According to Kofele-Kale[59] each generation is party to a duty 'to pass on to succeeding generations a planet at least as healthy as the one it inherited so that each generation will be able to enjoy its fruits'. The assumption of enforceable obligations is central here yet as in the case of all rights and expectations of co-existence, tolerance and respect, these can be hard to guarantee. This is so in citizenship relationships existing on a local basis and will be even more the case when trying to invoke new obligations. Nonetheless, it would be good if we could begin to characterise the twenty-first century as one in which we strive to preserve both human rights and human security but also recognise that such goals cannot be fully realised unless we demonstrate similar regard for environmental rights and environmental security—both now and into the future that subsequent generations will inherit.

[59] N Kofele-Kale, *The International Law of Responsibility for Economic Crimes* (Aldershot, Ashgate, 2006) 324.

problems arising from these are not necessarily immediate, and it can take a very long time for them to be realised. Because of this, a causal link between these events and acts of States is difficult to establish. It is also the case that the effects of climate change are by and large extraterritorial, and it is difficult to hold a particular State to account for something which happens in a far corner of the world. All of these pose an important question on the usefulness of a human rights framework in addressing the phenomenon.

The purpose of this chapter is to argue that, despite these practical difficulties in applying the human rights norms and principles, a human rights framework can still usefully supplement the existing framework on climate change as represented by the United Nations Framework Convention on Climate Change 1992[4] and the Kyoto Protocol 1997.[5] The chapter starts by exploring some of the key benefits of using a human rights framework, such as the promotion of a victim-centred and a holistic approach to the phenomenon as well as the involvement of independent bodies for scrutiny and monitoring of obligations. It continues with an examination of applicable human rights norms and principles through an analysis of the existing jurisprudence at regional and international levels. The relevant human rights, such as the rights to health, life, adequate standard of living, housing, property and healthy environment will be considered in depth. The final part of this chapter highlights some of the challenges of using the human rights framework, such as how to deal with acts of non-State actors, the difficulty in establishing a causal link and the extraterritorial application of human rights law. It will be argued that an obligation to protect human rights from interference by non-State actors, the precautionary principle (prevention of possible harms) and intergenerational responsibility can mitigate these obstacles, and therefore that the human rights framework can still play an important role in addressing climate change.

BENEFITS OF A HUMAN RIGHTS FRAMEWORK

There are several core benefits of utilising a human rights framework to address climate change, and it will be shown that they can usefully supplement the existing framework as represented by the UN Framework Convention and the Kyoto Protocol. First and foremost, a human rights framework puts individual human beings at the centre of any action against climate change. They are regarded as potential or actual victims of human rights violations, and the framework allows us to perceive climate change from the point of view of human dignity, with the recognition that it does have actual human costs, rather than treating it as a technical/scientific issue. As a result, protection of the environment does not become an end in itself, and it makes us realise that

[4] 1771 UNTS 10.

[5] 37 ILM 22 (1998).

there are wider issues which must be taken into consideration to protect the wellbeing of each individual. This can put additional moral as well as legal pressure on States to act, thereby supplementing the current legal framework, which has experienced some difficulties in relation to implementation.

Second, related to the first point, international human rights law offers avenues for those affected by climate change to seek a remedy. The UN Framework Convention is an instrument which obliges States to take preventive measures in order to mitigate the adverse effects of climate change. There is no doubt that this is an important first step, but the instrument itself does not address the consequences of climate change after individuals are victimised one way or another. As will be shown below, international human rights law does contain this element of preventing the future violations of human rights. However, it goes further than international environmental law by requiring States to provide remedies if relevant human rights are violated. Article 2(3) of the International Covenant on Civil and Political Rights 1966 (ICCPR),[6] for instance, provides that States Parties are under an obligation to 'ensure that any person whose rights or freedoms as herein recognized are violated shall have an effective remedy' which is to be determined by competent judicial, administrative or legislative authorities. Similar provisions exist in the European Convention on Human Rights 1950 (ECHR)[7] and the American Convention on Human Rights 1969 (ACHR).[8] Unfortunately, a comparable provision is not found in the UN Framework Convention and therefore victims of violations of that Convention can be left without an effective remedy.

Third, a human rights framework facilitates the involvement of independent mechanisms to monitor implementation of various obligations. At the international level, there are treaty bodies such as the Human Rights Committee and the Committee on Economic, Social and Cultural Rights, as well as special procedures including the Special Rapporteurs on the Rights to Health, Adequate Housing, Food and Water. As will be shown below, many of these bodies have taken up the issue of climate change directly or indirectly. At the regional level, there are judicial mechanisms such as the European and Inter-American Courts of Human Rights and the African Court on Human and Peoples' Rights. The key benefits of these independent bodies are twofold. First, they can provide an objective assessment of the implementation of various obligations established under relevant treaties without undue political interference. Second, they can elaborate and develop applicable human rights norms and principles which in turn will serve as guidance for States and other actors concerned with reducing the adverse effects of climate change.

[6] 999 UNTS 171.
[7] Art 13, ETS No 5 (1950).
[8] Art 25, 1144 UNTS 123.

Compared to the human rights regime, international environmental law has been slow in introducing judicial and other mechanisms of accountability.[9] A body established under the UN Framework Convention, the Conference of Parties,[10] is a political body which consists of State Parties, as the name suggests. This makes political interference more likely than within the human rights regime. While the Conference is obliged to monitor implementation of various obligations, States Parties are not likely to blame others strongly for the fear that they could face a similar consequence. There are subsidiary bodies which provide scientific and technical advice to the Conference and monitor implementation of the UN Framework Convention. However, they too consist of government representatives, as stipulated under Articles 9 and 10. It should be noted further that there are some expert bodies under the Framework Convention. They are the Consultative Group of Experts on National Communications,[11] the Least Developed Countries Expert Group[12] and the Expert Group on Technology Transfer.[13] These bodies, however, are to provide technical advice and are not charged with monitoring implementation of the legal framework on climate change.

A critique can of course be raised against the human rights bodies as it is common knowledge that enforcement of human rights norms and principles is regarded as weak. For several reasons, however, this should not be the reason why international human rights law and bodies should not be involved in dealing with climate change. First, as already indicated, enforcement of international environmental law is weaker than its human rights counterpart. While the human rights regime has judicial (regional courts) or quasi-judicial (treaty-based) bodies, many of which have been proactive in securing compliance, the latter does not have this component. Therefore, the human rights law regime can still augment international environmental law. Second, despite their weaknesses, human rights bodies have been able to provide objective assessment of States' compliance, facilitate transparency and mobilise public support by issuing concluding observations, for instance. Being a political body, there is no guarantee that the Conference of Parties under the UN Framework Convention will be able to carry out the same level of assessment as the human rights bodies. Third, independent human rights bodies can develop and articulate human rights norms and principles without undue political interference through individual communications and general comments. While the periodic reports submitted to the Conference of Parties

[9] International Council on Human Rights Policy, *Human Rights and Climate Change: A Rough Guide* (2008) 7.

[10] Art 7 of the UN Framework Convention.

[11] Decision 8/CP.5 (4 November 1999).

[12] Established by the Marrakesh Accords (2001).

[13] Decision 4/CP.7 (10 November 2001).

under the UN Framework Convention can play this role to some extent, the interpretation of norms and principles will always have political dimensions.

Finally, it is submitted that a human rights framework can promote a holistic approach to climate change. The current environmental law framework is mainly concerned with 'stabilising greenhouse gas concentration at a level that would prevent dangerous anthropogenic interference with the climate system.'[14] While this is one of the important steps to be taken, the framework in its current form is not capable of addressing wider issues relating to climate change. For instance, any effective action against climate change must be able to address the root causes of the emission of greenhouse gases. It is no secret that States, particularly from poorer regions of the world, engage in activities which cause climate change primarily for economic development. Economic development is a human rights issue, as clearly recognised in relevant human rights instruments including the International Covenant on Economic, Social and Cultural Rights 1966 (ICESCR),[15] the ICCPR,[16] and the Convention on the Elimination of Discrimination Against Women 1979 (CEDAW).[17] Therefore, this issue should form an important part of overall action against climate change under the human rights framework. While the UN Framework Convention does touch upon sustainable development in order to mitigate the effect of climate change,[18] it is not suited to addressing other pertinent issues such as long-term health effect of climate change, use of lands and natural resources by indigenous populations, natural disasters, forced displacement, and participation of citizens in environmental decision-making. This shortfall can be ameliorated by international human rights law as it recognises these and other wider issues relating to climate change as will be shown below. In summary, it seems reasonable to argue that international human rights law can usefully supplement international environmental law to address climate change.

RELEVANT HUMAN RIGHTS ISSUES

Right to Health

One of the most obvious human rights issues is the potential and actual impact of climate change on one's health. There is a wide variety of pertinent issues such as malnutrition and spread of tropical/infectious diseases (eg malaria and yellow fever) through natural disasters (eg flood and drought) which are caused directly or indirectly by climate change. International human rights law

[14] Art 2 of the Framework Convention.

[15] Arts 1 (Self-determination) 6 (Work) and 11 (Adequate Standard of Living), 993 UNTS 3.

[16] Art 1 (Right to Self-Determination).

[17] Art 14 (Rural Development), 1249 UNTS 13.

[18] Preamble, Arts 2, 3, and 4.

recognises the link between environmental degradation and one's enjoyment of health. The ICESCR is a good example. Article 12(2)(b) provides for the improvement of all aspects of environmentaland industrial hygiene. In examining this provision, the Committee on Economic, Social and Cultural Rights, an independent body charged with monitoring the implementation of the ICESCR, has stated clearly that the right to health embraces a wide range of socio-economic factors including underlying determinants of health such as a healthy environment.[19] In its view, the above-mentioned Article relates to prevention or reduction of exposure to harmful substances and other environmental conditions that directly or indirectly impact upon health.[20] A point to note here is the 'indirect impact' of environmental conditions. While the Committee does not mention climate change specifically, the emission of greenhouse gases, global warming and the spread of infectious and tropical diseases can qualify under this heading. Aside from the ICESCR, the Convention on the Rights of the Child 1989 (CRC)[21] also recognises the link between the right to health and the risk of environmental pollution under Article 24(2)(c).

In terms of State obligations, while the ICESCR does not impose obligations specific to climate change, there are general obligations which can be imposed to protect the right to health and the environment. As noted above, States are obliged to adopt preventive measures against exposure to harmful substances such as radiation, harmful chemicals or other detrimental environmental conditions. In addition, they should formulate and implement national policies aimed at reducing and eliminating pollution of air, water and soil.[22] The policy development relating to infectious or tropical diseases and their possible health consequences can fall into this category. When individuals are affected, States are also obliged to provide adequate and accessible health facilities such as hospitals.[23] Further, they have to adopt legislative and other measures to prevent non-State third parties from infringing one's right to health.[24] In this regard, the Committee on Economic, Social and Cultural Rights observes that a violation of the right to health can occur when States fail to enact laws to prevent the pollution of water, air and soil by extractive and manufacturing industries.[25] In the context of climate change, this reasoning could apply when the relevant industries conduct their business without due regard to the health of local communities and people.

[19] General Comment No 14 (The Right to the Highest Attainable Standard of Health) (2000), 'Compilation of General Comments and General Recommendations Adopted by Human Rights Treaty Bodies' HRI/GEN/1/Rev.7 (24 May 2004), para 4.

[20] Ibid, para 15.

[21] 1577 UNTS 3.

[22] Above, n 19, para 35.

[23] Ibid, para 12.

[24] Ibid, para 51.

[25] Ibid.

At the regional level, the negative impact of environmental degradation on the right to health has been recognised at least in the African context. The pertinent jurisprudence is the often cited case of *The Social and Economic Rights Action Center and the Center for Economic and Social Rights v Nigeria*[26] before the African Commission on Human and Peoples' Rights. This case concerned allegations that an oil consortium exploited oil reserves in Ogoniland without due regard to the health and the environment of local communities and that the government of Nigeria protected various activities of the consortium. The African Commission, in referring to Article 12(2) of the ICESCR, recognised the link between the environment and the right to health by emphasising that environmental degradations are harmful to physical and mental health.[27] In terms of States' obligations, the Commission stated that they were obliged to undertake appropriate monitoring of the environment, conducting a scientific study before any developmental activities, and providing meaningful opportunities for individuals to be heard and to participate in decision-making.[28] This case is relevant to climate change if the industrial activities which produce greenhouse gases result in air, soil or water contamination. It also applies to the manners in which these activities are carried out. In particular, if no adequate study and consultation are conducted prior to any industrial activity, this can be regarded as a violation of the human rights of local populations.

A similar case was also decided in Europe. In *Marangopoulos Foundation for Human Rights v Greece*,[29] the European Committee of Social Rights, which monitors implementation of the European Social Charter,[30] examined an issue of air pollution caused by lignite mining and combustion. While accepting that Greece had a legitimate objective of energy independence in relying on mining and use of lignite,[31] the European Committee recognised that lignite mining and combustion (particularly generation of nitrogen oxide, which contributes to CO_2 emission, and sulphur dioxide) could lead to serious problems for human health such as respiratory and cardiovascular diseases as well as cancer.[32] In the end, the Committee found Greece to be in breach of Article 11 of the European Social Charter on the right to health as it did not strike a reasonable balance between the interest of those affected and the general interest in energy independence.[33] In terms of obligations, the European Committee, like other bodies mentioned above, noted that States are obliged to inform and educate the general public,

[26] Communication No 155/96, ACHPR/COMM/A044/1 (27 May 2002).
[27] Ibid, paras 51 and 52.
[28] Ibid, para 53.
[29] Complaint No 30/2005 (6 December 2006).
[30] Adopted in 1961 and revised in 1996. ETS Nos 35 and 163.
[31] Above n 29, para 198.
[32] Ibid, para 200.
[33] Ibid, para 221.

assess the health risk through epidemiological monitoring, and enter into fair and genuine consultation with those affected.[34] This is a significant case as it has set a precedent for examining States' obligations under the Kyoto Protocol from a human rights perspective.[35] Another important aspect is the Committee's adoption of 'a living instrument approach,' an established principle under the European Convention on Human Rights,[36] where the Charter is to be interpreted in light of the present-day conditions. In the present case, the Committee clearly stated that Article 11 included the right to a healthy environment.[37]

Right to Adequate Standard of Living

The right to an adequate standard of living is also relevant to climate change. There are several dimensions to this right. To begin with, this right covers the right to food and water as stipulated, among others, in Article 11 of the ICESCR. Climate change can have an adverse impact upon food production and increase the likelihood of hunger, particularly in the poorest regions of the world.[38] The United Nations Development Programme in this regard has estimated that as many as 600 million people could face malnutrition due to climate change.[39] In a similar vein, climatic conditions such as drought and flooding have a negative impact on clean and adequate water.[40] The Special Rapporteur on the Right to Health has noted in this respect that people have been forced to rely on polluted water sources, which in turn has resulted in the spread of waterborne and other diseases.[41] In addition, the right to an adequate standard of living also covers housing. This right is defined as a right to live somewhere in security, peace and dignity.[42] There is now good evidence to suggest that climate change can prevent individuals from enjoying this important right. Sea level rises, storms and floods can become problematic, particularly for those living in coastal areas.[43] In the Arctic region and in some island States, these

[34] Ibid, paras 203 and 207.

[35] M Trilsh, 'European Committee on Social Rights: The Right to a Healthy Environment' (2009) 7 *International Journal of Constitutional Law* 529.

[36] *Tyler v United Kingdom* (1978), Series A, No 26.

[37] Above n 29, para 195.

[38] Report of the High Commissioner, above n 2, para 26.

[39] UNDP, 'Human Development Report 2006, Beyond Scarcity: Power, Poverty and the Global Water Crisis' (2006). See also the Report of the Special Rapporteur on the Right to Food, A/HRC/7/5 (10 January 2008), para 51.

[40] Report of the High Commissioner, above n 2, para 29.

[41] Report of the Special Rapporteur on the Right of Everyone to the Enjoyment of the Highest Attainable Standard of Physical and Mental Health, A/62/214 (8 August 2007), para 101.

[42] General Comment No 4 (The Right to Adequate Housing) (1991), Compilation of General Comments, above n 19, para 7.

[43] Report of the High Commissioner, above n 2, para 36.

natural disasters have resulted in forced displacement and relocation of peoples and communities.[44] Another important aspect of climate change is rural-to-urban migration facilitated as a result of erosion of livelihoods.[45] This leads to overpopulation in urban areas, which triggers shortage in housing, forcing people from rural areas to hazardous areas.[46]

There are various obligations imposed upon States in relation to these sets of rights. In relation to the right to water, States are under an obligation to adopt strategies and programmes to ensure that there is sufficient and safe water for present and future generations, which include assessing the impacts of actions that may impinge upon water availability and natural-ecosystem watersheds, such as climate changes, desertification, increased soil salinity, deforestation and loss of biodiversity.[47] The specific mention of climate change is significant here. States also have to refrain from actions that interfere, directly or indirectly, with the enjoyment of the right to water in other countries.[48] This means that any activities undertaken within the State party's jurisdiction should not deprive other States of their ability to realise the right to water for persons in their jurisdiction.[49] Steps should also be taken by State parties to prevent their own citizens and companies from violating the right to water of individuals and communities in other countries.[50] In terms of the right to adequate housing, disaster reduction measures such as conducting risk assessment in urban planning, improving building infrastructure, and setting up alarm and evacuation systems, are all regarded as necessary.[51]

Right to Life

Another related and important right is the right to life. In addition to infectious and tropical diseases, people can lose their life through heat waves, floods, storms, drought and other natural disasters exacerbated by climate change.[52] There is well-established jurisprudence, at least in Europe, that the right to life

[44] Ibid.

[45] Ibid, para 37.

[46] Ibid.

[47] General Comment No 15 (The Right to Water) (2002), Compilation of General Comments, above n 19, para 28.

[48] Ibid, para 30. A similar reasoning has been adopted in relation to the right to food. See General Comment No 12 (The Right to Food) (1999), *Compilation of General Comments*, above n 19, para 36.

[49] Ibid.

[50] Ibid, para 33. Similar obligations are recognised in relation to the right to food. General Comment No 12, above n 48, para 36.

[51] Report of the Special Rapporteur on the Right to Adequate Housing, A/64/255 (6 August 2009) paras 51–54.

[52] Report of the High Commissioner, above n 2, paras 22–23.

can be engaged if States fail to regulate environmental risks.[53] In *Oneryildiz v Turkey*, the European Court of Human Rights held that Article 2 (right to life) is applicable to any activity, public or private, where the right is at stake.[54] In the context of industrial activities, the European Court stated that dangerous activities are pertinent.[55] In this case, the operation of a waste management site was regarded as such. Applying this reasoning to climate change, if industrial activities which produce greenhouse gases are conducted in a way as to endanger the life of those living nearby, then the right to life can surely be engaged. This reasoning is quite similar to the ones adopted by other human rights bodies noted above. Further, it was held in *Budayeva and Others v Russia* that the right to life can also be relevant to natural disasters which are beyond States' control.[56] A note of caution, however, is that a natural disaster must be imminent and identifiable,[57] and this does make it difficult for disasters resulting from climate change as its effects are normally realised in the future. In terms of obligation, in addition to a positive obligation to inform the general public about life-threatening events,[58] it has been held that States are also sometimes under a positive duty to institute a judicial inquiry into deaths caused and to take preventive measures to mitigate natural hazards.[59]

The Rights to Private and Family Life and Property

In addition to obvious rights such as the rights to health and life, it has been recognised that environmental damage and harm can affect the right to private and family life and right to home in particular. One of the often cited cases is *Lopez Ostra v Spain* before the European Court of Human Rights. In this case, the European Court recognised, among others, that severe environmental harm (pollution) can affect individuals' well-being and prevent them from enjoying their right to private and family life without endangering their health.[60] The basis of this decision was that the prolonged nuisances caused by a waste management plant prevented the applicant from enjoying her right to respect for her home. This decision therefore is closely inter-linked to adequate housing, as noted above. A note of caution, however, is that the mere existence of environmental degradation is not

[53] A Boyle, 'Human Rights or Environmental Rights? A Reassessment' (2007) 18 *Fordham Environmental Law Review* 471, 487.

[54] Application no 48939/99, Judgment of 30 November 2004, para 71.

[55] Ibid.

[56] Applications nos 15339/02, 21166/02, 20058/02, 11673/02 and 15343/02, Judgment of 29 September 2008, paras 128–37. This case was about a mudslide killing several people and destroying many buildings.

[57] Ibid, para 137.

[58] Oneryildiz, above n 54, para 90.

[59] Budayeva, above n 56, paras 131 and 137.

[60] Application no 16798/90, Judgment of 9 December 1994, para 51.

sufficient to hold States accountable. In this respect, in *Fadeyeva v Russia*, the European Court of Human Rights stated that the adverse effect of environmental degradation such as pollution must attain a minimum level.[61] Determination of this 'minimum level' depends on all the circumstances, including the intensity and duration, as well as the physical and mental effects of a nuisance in question.[62] Therefore, it is extremely difficult to provide general guidance applicable to all situations, and this equally applies to climate change.

In terms of States' obligations, it has been held that access to information and the right to participate in decision making must be secured.[63] While jurisprudence noted above already touched upon these, they are also recognised under Article 25 of the ICCPR, which provides for the right to take part in public affairs. The term 'public affairs' is to be interpreted broadly to encompass the exercise of legislative, executive and administrative powers and covers all aspects of public administration, and the formulation and implementation of policy at international, national, regional and local levels.[64] Decision making relating to sustainable development, environmental protection and climate change undoubtedly fall under this heading. It is also important to stress that this right to participate in public affairs is interlinked with other human rights such as the rights to freedom of expression and assembly and association.[65]

A related right to private and family life is the right to property. The relevance of this to climate change would be apparent in the exploitation of lands and natural resources which leads to emission of greenhouse gasses. In *Maya Indigenous Community of the Toledo District v Belize*, before the Inter-American Commission on Human Rights, it was argued that granting logging and oil concessions on lands used and occupied by the Maya community violated, among others, their right to property under Article XXIII of the American Declaration of Human Rights.[66] The Inter-American Commission held that 'the right to use and enjoy property may be impeded when the State itself, or third parties acting with the acquiescence or tolerance of the State, affect the existence, value, use or enjoyment of that property without due consideration of and informed consultations with

[61] Application no 55723/00, Judgment of 30 November 2005, para 69.

[62] Ibid.

[63] *Guerra and others v Italy*, Application No 14967/89, Judgment of 19 February 1998 and *Taskin and others v Turkey*, Application No 46117/9, 10 November 2004.

[64] General Comment No 25 (The Right to Participate in Public Affairs) (1996), Compilation of General Comments, above n 19, para 5.

[65] Ibid, para 8.

[66] Case 12.053, Report No 40/04, Inter-Am. C.H.R., OEA/Ser.L/V/II.122 Doc 5 rev 1 at 727 (2004), para 46. Article XXIII reads: '(e)very person has a right to own such private property as meets the essential needs of decent living and helps to maintain the dignity of the individual and of the home.'

of the African Commission gains widespread acceptance in the future, this will be advantageous for climate change as it can be addressed under the heading of 'the right to environment' without the need to rely on other related rights such as the right to health and an adequate standard of living. The drawback, however, is that the right to environment is recognised only in certain regions of the world, and, in the absence of a universal instrument, its relevance in other regions such as Asia and the Middle East will be called into question.

KEY CHALLENGES

The above analysis clearly demonstrates that climate change engages a wide variety of human rights issues and that certain obligations can be imposed upon States. However, there are some key challenges which must be overcome in order for international human rights law to be truly useful in addressing climate change. The first challenge is the fact that international human rights law bind States and therefore non-State actors are not held directly accountable. This poses a dilemma for climate change, as most activities leading to the emission of greenhouse gases are conducted by private and non-State industries. The Human Rights Committee makes this clear by stating that 'obligations are directed to States and do not, as such, have direct horizontal effect as a matter of international law.'[84] Nevertheless, international human rights law is still relevant for these actors as it still imposes an obligation on States to prevent them from violating human rights. This is inherent in 'an obligation to protect,' which has been interpreted as binding States to prevent third parties from interfering with human rights.[85] The jurisprudence and opinions of the human rights bodies noted above also support this[86] and therefore this obligation is relevant for climate change.

The second challenge is identification of a State responsible for harms caused as a result of, or related to, climate change. At issue here is the need for a causal link between an act of a particular State and the effect on climate change and human rights. If the harm is immediate, then it is relatively easy to establish responsibility, as identification of victims is somewhat straightforward. However, the effects of climate change are far from immediate. A question may be asked, for instance, as to which State should be held liable for the recent flood in Australia. If Australia has failed to take any reasonable preventive measures, it could be held liable, as the jurisprudence above

[84] General Comment No 31 (The Nature of the General Legal Obligation Imposed on States Parties to the Covenant) (2004), *Compilation of General Comments*, above n 19, para 8.

[85] *The Maastricht Guideline on Violations of Economic, Social and Cultural Rights*, para 6, reprinted in (1998) 20 *Human Rights Quarterly* 691.

[86] General Comment No 14, above n 19, para 51, and General Comment No 15, above n 47, para 33.

demonstrates. What if climate change is partly responsible for this? How can we trace the source of the problem and attach responsibility to a particular State? Also, multiple States could be responsible for climate change, and the flood could have been caused by a number of factors. The human rights jurisprudence does support the principle that a particular harm must be immediate. The Human Rights Committee, for instance, held in *Bordes and Temeharo v France*[87] that the risk of harm from nuclear radiation resulting from nuclear testing was too remote for the claimants to qualify as victims.

Another related problem is the extra-territorial application of international law generally and international human rights law in particular. In an often cited case of *Trail Smelter Arbitral Decision*,[88] an arbitral tribunal established that:

> under the principles of international law, as well as of the law of the United States, no State has the right to use or permit the use of its territory in such a manner as to cause injury by fumes in or to the territory of another or the properties or persons therein, when the case is of serious consequence and the injury is established by clear and convincing evidence.

This decision recognised State responsibility for transboundary harm, which has become one of the important principles in international environmental law. This principle is clearly recognised in relation to climate change, as can be seen in the Preamble of the UN Framework Convention, which provides that 'States have… the responsibility to ensure that that activities within their jurisdiction or control do not cause damage to the environment of other States or of areas beyond the limits of national jurisdiction.'

However, the application of the principle developed in *Trail Smelter* is controversial for international human rights law, because its extraterritorial application is limited to certain circumstances. As a general rule, international human rights law protects individuals within the jurisdiction of a particular State. This is affirmed in Article 2 of the ICCPR, Article 1 of the ECHR and Article 1 of the ACHR. In terms of climate change, it would be easy to establish responsibility if industrial activities within a given State violate the human rights of those residing in that State, as in the *Ogoniland Case*. What happens if industrial activities within a State violate the human rights of individuals in another State? In an often cited case of *Bankovic v Belgium*, the European Court of Human Rights held that an obligation arises where a State is in effective control of a particular territory in a sense of exercising some sort of public powers.[89] This reasoning is affirmed by the Human Rights Committee.[90] Applying this to climate change, a State

[87] Communication No 645/1995, CCPR/C/57/D/645/1995.
[88] (1939) 33 *American Journal of International Law* 182.
[89] (2001) BHRC 435, para 47.
[90] General Comment No 31, above n 84, para 10.

could be held liable if it conducts industrial activities in a territory under its control (eg military control). However, such an instance is not very common, and the jurisprudence is vague in relation to cases where activities violate the human rights of those living in another State or territory not under any form of control. An added complexity is that the reasoning of the European Court in *Bankovic* applies to exercise of 'public' powers—ie acts of State agents abroad. This therefore excludes the activities of non-State actors.[91]

Do all of these points mean that international human rights law is irrelevant to climate change? It is submitted that it still has a role to play. To begin with, States have a general obligation to prevent future or possible violations of human rights. A useful concept in understanding this further is the precautionary principle. Simply put, it provides that where there are threats of serious or irreversible damage, a lack of full scientific certainty should not be used as a reason for postponing such measure. This principle is enshrined in Article 3(3) of the UN Framework Convention. In the context of human rights law, this principle was supported in *Tatar v Romania*[92] where the European Court of Human Rights held that a State must adopt measures against serious risks to health and well-being even in the absence of concrete evidence of such risks. In so doing, the European Court recognised that the precautionary principle has transformed itself from a philosophical concept into a judicial norm, at least in the European context.[93] It is also important to stress here that the Court was referring to transboundary harm as the action of Romania affected neighbouring Serbia and Hungary. Similar reasoning was adopted by the European Committee on Social Rights when it stated that States have an obligation to prevent air pollution not only at the local level, but also to help reduce it on a global scale.[94] What becomes apparent here is that the precautionary principle and the duty to prevent potential harms can mitigate the problems arising from a lack of clear causal link and the extraterritorial application of international human rights law.

Another point to note is that, while the extraterritorial obligation in relation to civil and political rights might be limited as highlighted in *Bankovic*, a different picture emerges in relation to economic and social rights. As the ICESCR does not refer to territory or jurisdiction as the criteria for its scope and application, it has been argued that a certain

[91] N Vennemann, 'Application of International Human Rights Conventions to Transboundary State Acts' in RM Bratspies and RA Millar (eds), *Transboundary Harm in International Law: Lessons from the Trail Smelter Arbitration* (Cambridge, Cambridge University Press, 2006) 306.

[92] Application No 67021/01, Judgment of 27 January 2009.

[93] D Shelton, 'Developing Substantive Environmental Rights' (2010) 1 *Journal of Human Rights and the Environment* 89, 107.

[94] Above n 29, para 203.

extraterritorial scope was intended by drafters.[95] Indeed, the extraterritorial application of economic, social and cultural rights has been acknowledged by human rights bodies. In examining the right to health, the Committee on Economic, Social and Cultural Rights clearly stated that 'States parties have to respect the enjoyment of the right to health in other countries.[96] The Special Rapporteur on the Right to Food has also expressed a similar opinion.[97] Interestingly, he continued to state that States have to prevent their own citizens, companies and other third parties subject to their jurisdiction from violating the right to food in other States. A point to stress here is that States have to address not only their own acts, but also those of non-State actors.[98] What becomes apparent, then, is that the limitation imposed upon civil and political rights in terms of their extraterritorial application can be ameliorated by economic and social rights, all of which are closely interlinked with one another as the analysis above clearly demonstrates.

A danger of relying too heavily on economic, social and cultural rights is that the nature of obligations is different from civil and political rights. While civil and political rights are generally regarded as imposing obligations of conduct or immediate effect, economic, social and cultural rights are to be realised progressively over the course of time,[99] and this is explicitly recognised in Article 2(1) of the ICESCR. The progressive realisation is acknowledged because it is difficult for States, particularly those in poorer regions of the world, to realise these sets of rights due to a lack of adequate resources and means. This, however, should not be the reason for ignoring or delaying implementation of relevant obligations, as the ICESCR simultaneously provides for international co-operation and assistance under the same Article. This is further augmented by Articles 1, 55 and 56 of the United Nations Charter.[100] While all States should be able to benefit from

[95] F Coomans, 'Some Remarks on the Extraterritorial Application of the International Covenant on Economic, Social and Cultural Rights' in F Coomans and M Kamminga (eds), *Extraterritorial Application of Human Rights Treaties* (Antwerp, Intersentia, 2004) 185. See also M Gondek, *The Reach of Human Rights in a Globalising World: Extraterritorial Application of Human Rights Treaties* (Antwerp, Intersentia, 2009) 298–99.

[96] General Comment No 14, above n 19, para 39. See also General Comment No 15, above n 47, para 30, in relation to the right to water.

[97] E/CN.4/2005/47 (24 January 2005), para 48.

[98] See also General Comment No 14, above n 19, where the Committee on Economic, Social and Cultural Rights stated, at para 39, that States must prevent third parties from violating the right in other countries, if they are able to influence these third parties by way of legal or political means.

[99] General Comment No 3 (The Nature of State Parties Obligations) (1990), *Compilation of General Comments*, above n 19, para 1.

[100] These provisions touch upon international co-operation in solving international economic, social, cultural, health or humanitarian problems, as well as in promoting universal respect for, and observance of, human rights and fundamental freedoms.

international co-operation and assistance,[101] developing States are likely to be the main recipients of such co-operation and assistance.

An important question which should be asked here is whether developed States are under an obligation to assist developing ones. *Travaux préparatoires* of the ICESCR suggests that while developing States are entitled to request assistance, it is not a legal right.[102] The developed States also consistently deny the existence of any clear obligation to assist them.[103] Increasingly, however, it has been argued by scholars that such a position is untenable in the twenty-first century,[104] and the Committee on Economic, Social and Cultural Rights has interpreted various rights as entailing a duty to assist developing States. In relation to the right to water, for instance, the Committee noted:

> International assistance should be provided in a manner that is consistent with the Covenant and other human rights standards, and sustainable and culturally appropriate. The economically developed States parties have a special responsibility and interest to assist the poorer developing States in this regard.[105]

Craven, one of the foremost authorities on economic, social and cultural rights, also states that while developed States are not required to meet all needs of developing States, they are under a duty to provide some form of assistance.[106] It may therefore be argued that developed States should at least take a lead in assisting others to realise economic, social and cultural rights.

It is worth noting here that the obligation to assist developing States under the ICESCR is supplemented by the UN Framework Convention, through the notion of 'common but differentiated responsibilities and capabilities.' This is stipulated under Article 3(1) and acknowledges that, although all States are under an obligation to combat climate change, responsibilities are bound to be different due to inequalities and varied capacities which exist among them. In particular, developed States, which have been mainly responsible for climate change, should take a lead in combating the phenomenon.[107] In so doing, they should provide financial and other assistance to developing States so that the latter can fulfil obligations stipulated under the Framework Convention.[108] The provisions such as these are intended to 'progressively modify the existing status quo' in order to achieve fairness and equity among

[101] R Kunnemann, 'Extraterritorial Application of the International Covenatn on Economic, Social and Cultural Rights' in *Extraterritorial Application of Human Rights Treaties*, above n 95, 204.

[102] M Craven, *The International Covenant on Economic, Social and Cultural Rights: A Perspective on Its Development* (Oxford, Clarendon Press, 1998) 149.

[103] M Ssenyonjo, *Economic, Social and Cultural Rights in International Law* (Oxford, Hart Publishing, 2009) 77.

[104] Ibid.

[105] General Comment No 15, above n 47, para 34.

[106] Above n 102, 149.

[107] Art 3(1).

[108] Art 4.

States.[109] This resonates well with the progressive realisation of economic, social and cultural rights, as it is recognised that States can fulfil obligations at a different pace depending on available resources and capacity,[110] and further strengthens the argument for proactive co-operation and assistance to those in need.

Finally, the principle of intergenerational equity or responsibility is useful in overcoming practical difficulties noted above. This is particularly attractive in relation to climate change as its effects are often realised in the future. Based on this, an argument may be made that States have to respect and promote the human rights of not only the current generation, but also of future generations. While this principle is recognised in international environmental law, including the UN Framework Convention,[111] it is equally relevant to international human rights law.[112] For instance, the Preamble of the Universal Declaration of Human Rights speaks of the equal and inalienable rights of all members of humanity. Weiss argues that the phrase 'all members of humanity' is wide enough to include all generations, both present and future, in the human family.[113] The same wording can be found in legally binding instruments such as the ICCPR and the ICESCR.

The human rights bodies also recognise the protection of the human rights of the future generation. For instance, the Committee on Economic, Social and Cultural Rights has stated that 'the notion of sustainability is intrinsically linked to the notion of adequate food or food security, implying food being accessible for both present and future generations.'[114] In a similar vein, the Special Rapporteur on the Illicit Movement and Dumping of Toxic and Dangerous Products stated that States must take into account the future costs and long-term impact of environmental degradations, as well as their obligation to save future generations from a multitude of health problems.[115] This obligation to protect future generations and to prevent future harm demonstrates that international human rights law is forward-looking, as it is not concerned only with remedying violations already caused. Boyle has noted that human rights protection benefits only the victims of a violation of rights.[116] This is true from a procedural point of view, as only victims can bring claims of violations in most cases. The human rights analysis

[109] P Cullet, 'The Kyoto Protocol and Vulnerability: Human Rights and Equity Dimensions' in S Humphreys (ed), *Human Rights and Climate Change* (Cambridge, Cambridge University Press, 2010) 186–87.

[110] Ibid.

[111] Art 3(1) provides that '(t)he Parties should protect the climate system for the benefit of present and future generations of humankind on the basis of equity.'

[112] EB Weiss, 'Our Rights and Obligations to Future Generation for the Environment' (1990) 84 *American Journal of International Law* 198, 200.

[113] Ibid 201.

[114] Above n 48, para 7.

[115] A/HRC/7/21 (18 February 2008), para 69.

[116] Above n 53, 505.

of climate change, however, illustrates that international human rights law is also relevant for 'all members of humanity' and not just victims. In conclusion, there are ways to overcome some of these challenges posed to international human rights law, and therefore this branch of law can usefully supplement international environmental law as represented by the UN Framework Convention to facilitate action against climate change.

CONCLUSIONS

This chapter has analysed climate change from a human rights perspective. It is evident that the phenomenon raises a wide variety of human rights concerns, and some obligations are imposed upon States to address it. Although there are some limitations inherent in international human rights law, it was shown that this branch of international law offers an additional avenue to encourage States to act. The human rights analysis of climate change also leads to several other points. First, the human rights norms and principles relating to climate change are at a very early stage of development, as the human rights bodies have not been very proactive in paying attention to the matter and developing a normative framework through general comments, concluding observations or jurisprudence. However, it is apparent that the principles applicable to environmental protection generally are equally important to climate change. At the very minimum, therefore, States are expected to observe them. Second, a human rights analysis of climate change underscores one of the fundamental principles of international human rights law that all human rights are indivisible, interdependent and interrelated. It is not possible to speak of the right to private and family life without paying due regard to the rights to property, culture and adequate standard of living simultaneously. This requires all of those concerned to adopt a broader human rights framework to address climate change. Third, international environmental law and human rights law, seemingly different branches of international law, share strong similarities in practice. The precautionary principle, intergenerational responsibility, common and differentiated responsibility and capability are good examples of this. All of these suggest that both branches of international law are mutually reinforcing and can usefully supplement the current action to combat climate change. The phenomenon is complex and multi-dimensional and therefore requires a holistic and flexible approach. The analysis throughout this chapter has hopefully shown that international human rights law can be a step in the right direction to promote such an approach.

8

Climate Change and Aid Funding: An Appraisal of Recent Developments

ANNAMARIA LA CHIMIA*

INTRODUCTION

THIS CHAPTER IS dedicated to the aid element of climate change financing; it investigates the role played by development aid in addressing the negative effects of climate change and the implications that the climate change discourse could have upon development aid projects. The chapter has the twofold aim of classifying the different types of aid employed to financing climate change projects and to identify the various ways in which climate change is penetrating the development aid sector, ie how development aid is being influenced by climate change policies. When analysing the different typologies of aid employed to address climate change, particular attention will be paid to new ways of using development aid, such as the use of development procurement to boost environmental objectives, which have yet to receive due attention from development experts.

The classification proposed in this chapter has both a clarification and an innovative purpose. Indeed, while the first two definitions studied—namely, climate aid and development aid for climate change—have gathered a lot of attention by the doctrine in recent years, the final category—climate change as a corollary policy of development aid—explores other forms of usage of development aid for addressing climate change and environmental issues, which have not received considerable attention by the literature (this is especially so as regards donors' use of development aid procurement to boost climate change policies).

A critical review of the existing literature will be provided in a systematic way and with the scope of both clarifying and guiding the reader through the literature present to date. The problems with some of the definitions currently

*I am grateful to the British Academy for funding the project of which this chapter forms a part. I also would like to thank Richard Craven for valuable research assistance.

employed will be highlighted. This paper, by classifying the different ways in which development aid is used in climate change financing, will also help a full understanding of the role played by development aid in climate change financing, and will bring to light the problems that might be linked to the use of development aid in climate change financing.

It is hoped that, by organising the different aid categories in the way proposed in our classification, practitioners and NGOs will be able to call for greater transparency from policy makers and donors when disbursing aid money and will be able to verify whether the funding which is being disbursed to financing climate change is truly *new and additional* (ex Art 11 of the Protocol and Art 4(3) of the Convention) and not in substitution to funding already allocated to development aid (see below). The lack of clarity on what (and how) aid is being disbursed for addressing climate change risks undermining the possibility of verifying whether needs are being met and if they are being met in an efficient way.

The paper will start with an overview of the interaction between climate change and promoting development in poor countries. The following section will deal with the use of aid as a tool to address the negative effects of climate change. Finally section four will deal with the classification of the different aid categories employed for financing climate change, and section five concludes.

CLIMATE CHANGE AND DEVELOPMENT: AN OVERVIEW

The negative impact of climate change on poverty eradication and on the achievement of the Millennium Development Goals is now widely recognised. The Human Development Report warns that 'delayed action on climate change will condemn more and more families to poverty and expose the most vulnerable to ecological disaster.'[1] Oxfam has estimated that somewhere around '375 million people will be affected by humanitarian disasters related to climate change by 2015—a 50% rise over past years.'[2]

The poorest nations, living in lower latitudes (especially in Africa and Asia) will suffer the most from climate change-related catastrophes. Droughts, flooding, coastal erosions, water scarcity, will be some of the effects associated with climate change, and these, in turn, will hamper people's capacity to produce food, to secure shelter, to access land and safe-drinking water, constraining further the scarce resources of many developing countries already struggling to achieve development targets—such as food

[1] 'Why Coherence Counts for Development' in OECD Observer No 270/271, (December 2008–January 2009), available at www.oecdobserver.org.

[2] Reported in R Voss, 'Adaptation Aid: Dealing With Climate Change in the Developing World' (July 2009), available at www.aidemocracy.org/students/adaptation-aid-dealing-with-climate-change-in-the-developing-world/; see further Ganeshan and Diamond, 'Forecasting the Numbers of People Affected Annually by Natural Disasters up to 2015' Oxfam BB (April 2009).

security—and to fulfil their human rights obligations—such as the right to food, right to water, right to health, etc.

Furthermore, the negative effects of climate change are expected to worsen in the near future. The Intergovernmental Panel on Climate Change IPCC reports that by 2020 up to 250 million more people in Africa could face increased water scarcity and agricultural yields decline[3] with dramatic consequences on the capacities of those people to meet their development needs. Development practitioners are now very active in this field, as 'they have realised the potential implications of climate change for development.'[4]

Boyd *et al* have argued 'Climate change challenges development policy and planning processes in several important ways. Planning and visioning for development futures relate to scenarios of future possibilities in a warmer world. The challenge lies in how to plan for development futures, given the risk of runaway climate change, tipping points and huge uncertainty. Future development scenarios will depend upon the impacts of climate change.'[5]

Unfortunately, however, the recognition of the negative effects of climate change is yet to result in coherent and systematic strategies to prevent and tackle such negative consequences of climate change. Albeit the UNFCCC entrusts developed country Parties to the Convention to take the lead in combating climate change and to support developing country Parties in 'adapting to, as well as mitigating, climate changes'[6] little has been achieved so far. Many are the public and private financial instruments individuated to address climate change (such as proceeds from Clean Development Mechanism CDM) but a coherent and effective strategy to use them is lacking. As will be seen below, aid is amongst the instruments suggested to tackle climate change; however, its role is not clearly defined and the use made so far of aid money has been arbitrary and ambiguous.

FINANCING CLIMATE CHANGE: THE CASE OF AID FUNDS

Both the Convention and the Protocol foresee (commit in the case of the Protocol) rich countries to disburse financial assistance to developing

[3] Reported in L Peskett, N Grist, M Hedger, T Lennartz-Walker and I Scholz (hereafter Peskett *et al*), 'Climate change challenges for EU development co-operation: emerging issues', European Development Cooperation to 2020, Working Paper no 3 (January 2009).

[4] E Boyd, N Grist, S Juhola and V Nelson (hereafter Boyd *et al*), 'Exploring Development Futures in a Changing Climate: Frontiers for Development Policy and Practice' (2009) 27(6) *Development Policy Review* 659, 661.

[5] Boyd *et al*, ibid at 666. The UNFCCC also calls for the inclusion of climate change policies in development planning see Art 2 of the UNFCCC.

[6] UN Framework Convention on Climate Change, 'Investment and Financial Flows to Address Climate Change: An Update', Technical Paper, FCCC/TP/2008/7, (26 November 2008) available at unfccc.int/resource/docs/2008/tp/07.pdf, art 3, para 1.

countries to help them implement the Convention (Arts 3(1), 4(3), 4(4), 4(5), 12 of the Convention) and the Protocol (see Art 11 of the Kyoto Protocol which also recognises the need for the financial mechanisms to fund activities by developing country Parties).

True to an emerging notion of 'climate justice,' developed countries have recognised that the contribution of countries to climate change and their capacity to prevent and cope with its consequences vary enormously.[7] Industrialised countries, which have been responsible for the vast majority of greenhouse gas emissions that are causing the negative effects of climate change, have agreed to take the lead in meeting the IPPC recommendations of cutting greenhouse gas emissions and to help developing countries in mitigating and adapting to the effects of climate change.[8]

Developing countries need international assistance to support adaptation and mitigation projects (Art 4.4, 4.8 and 4.9 of the Convention).[9] International assistance will include funding, technology transfer and capacity building.[10] Addressing the negative effects of climate change will require conspicuous economic resources.[11]

'The world's richest countries have together pledged nearly $18 bn in the last seven years,' however, despite these commitments, less than $0.9 bn has been disbursed.[12] According to the UN Development Program investments of $86 billion per year are necessary to deal with climate change-related crisis. Such an estimate is reinforced by developing countries who believe that the amounts needed to help them adapt to the challenges of climate change vary between US$50 and US$100 billion per year.[13]

Many also denounce the inadequacies, not only in terms of quantity but also quality, of current formal financial mechanisms for responding to climate change.[14]

Aid is expected to also play a central role in addressing climate change. Aid will be used to fund capacity building projects and as part of other

[7] See unfccc.int/cooperation_and_support/financial_mechanism/items/2807.php.

[8] IIED (2004) 3; reported in C Saunders, 'The Stop Climate Change Chaos Coalition: Climate Change As a Development Issue' (2008) 29(8) *Third World Quarterly* 1509, 1510.

[9] UNFCC text available at unfccc.int/essential_background/convention/background/items/1362.php.

[10] See UNFCC website available at unfccc.int/cooperation_and_support/technology/items/1126.php; and unfccc.int/cooperation_and_support/capacity_building/items/1033.php.

[11] P Newell, J Timmons Roberts, E Boyd and S ul Huq (hereafter Newell *et al*), 'Billions at Stake in Climate Finance: Four Key Lessons' International Institute for Environment and Development Briefing (November 2009), available at http://pubs.iied.org/pdfs/17075IIED.pdf.

[12] Newell *et al*, ibid.

[13] United Nations Development Program, 'Human Development Report 2007/08: Fighting Climate Change: Human Solidarity in a Divided World' (Palgrave Macmillan, 2007) available at hdr.undp.org/en/media/HDR_20072008_EN_Complete.pdf.

[14] S Huq and JM Ayers, *Community Based Adaptation to Climate Change: An Update* (London, International Institute for Environment and Development, 2009) cited in Boyd *et al*, above n 4.

funding mechanisms in support of adaptation and mitigation projects.[15] The OECD reports that in 2007, Development Assistance Committee (DAC) members allocated approximately USD 3.8 billion for climate-change-related aid (where climate change is marked as 'principal' or 'significant' objective, see below). Nineteen DAC members reported a total of 4504 individual 'climate-change-related aid' activities. Aid money devolved to climate change is expected to grow in the future.

The notice of 'climate justice'—would be at the heart of a moral (but some sustain also legal) case for financing adaptation and mitigation projects through aid.[16] This is because those who are 'least responsible for creating climate change are suffering the most. Rich countries which became wealthy by burning fossil fuels, have a responsibility to assist and the capacity to do so.'[17]

Many believe that developed countries' responsibility to aid developing countries goes beyond a mere moral obligation.[18] The notion of 'climate justice' associated with the principles of 'common but differentiated responsibilities' and the 'polluters pay' principles – which find their legal base in Article 3, paragraph 1, of the UNFCCC and Art 11 of the Protocol – would justify the existence of a legal (not just a moral) obligation to grant climate aid.[19] Indeed, according to Article 3(1) of the UNFCCC developed countries Party to the Convention are expected to lead the fight against climate change and to *support* developing country Parties in 'adapting to, as well as mitigating, climate changes'[20] (emphasis added.)

If the existence of a legal responsibility to grant aid was recognised by the international community it would constitute a real breakthrough in international aid policies. Indeed, countries, donors of aid, have always regarded the act of granting aid a moral rather than a legal obligation and they have refused to recognise the existence of any international law obligation behind their aid activities (this is the case whether aid is disbursed to face humanitarian catastrophes or to address development issues).

[15] However, aid money is only one of the components of the funding mechanisms for adaptation and mitigation projects. For example the Adaptation Fund which was established to finance adaptation projects and programmes in developing countries that are Parties to the Kyoto Protocol is financed mainly with a share of proceeds from CDM project activities. Government contributions which qualify as aid money only constitute a small part of the Fund. See UNFCC website available at unfccc.int/cooperation_and_support/financial_mechanism/adaptation_fund/items/3659.php.

[16] Newell *et al*, above n 11.

[17] Oxfam, 'Beyond Aid: Ensuring Adaptation to Climate Change Works for the Poor' 132 (16 September 2009) Oxfam Briefing Note, available at www.oxfam.org.uk/resources/policy/climate_change/downloads/bp_132_beyond_aid_150909_web.pdf.

[18] See Voss, above n 2 and also Newell *et al*, above n 11.

[19] See Voss, above n 2.

[20] UN Framework Convention on Climate Change, above n 6.

The prevailing view is that the existing international law instruments, including Human Rights Covenants and Agreements such as the International Covenant on Economic Social and Cultural Rights or the Declaration on the Right to Development do not bind donor countries to grant aid. Seldom has it been sustained that developed countries, donors of aid, have a legal obligation to grant aid. Even when such attempts have been made, they have proved unsuccessful because of a multiplicity of factors, such as the open refusal of developed countries (signatory states to the above international Human Rights Agreements) to agree to any such obligation, or the vagueness of the wording of the obligation endorsed with such agreements which would not allow to implicitly ascribe any such obligation from the letter of the text, or because of the lack of enforceability of such instruments.

However, this might now be different in the case of climate aid given the presence of Articles 3 and 4 (see in particular Arts 4(3), 4(4) and 4(5)) of the UNFCCC and Article 11 of the Protocol. If a legal obligation to grant climate aid were recognised, there might be repercussions on the nature of the obligation to grant development aid. Unfortunately, it is still too early to attempt any such conclusion.

However, it seems that factors similar to those advocated against recognising a legal obligation to grant development aid could apply to climate aid.

In particular, to reject the view that there exists a legal obligation to grant climate aid one can adduce the lack of recognition of any such legal obligation by developed country signatories to the UNFCCC (funding remains voluntary under the UNFCCC), the lack of (or ineffectiveness of) enforcement mechanisms within the UNFCCC,[21] or even the fact that aid is not the only financial instrument used to tackle climate change (and to which Art 3 of the UNFCCC refers) and a communality of approaches for using the various financial instruments is lacking, rendering any commitment too vague to join the rank of legal obligation.

Denying the existence of a legal obligation to grant development aid, however, diminishes the importance of aid as a financial instrument to address climate change. Interestingly, one of the deal breakers in Copenhagen seems to have been a failed attempt to reach an agreement on the nature and quantity of aid to be devolved to the climate change cause.[22]

Although climate change funding has not yet reached the same level of disbursement of development aid, many believe that in the near future climate aid 'will equal or dwarf those of current development expenditure.'[23] Hence, climate change financing represents a gripping opportunity for developing countries. This is so especially at times of constrained budget

[21] See the paper presented by professor Geir Ulfstein at the conference 'Legal or Criminological Consequences of Climate Change' Oñati 2010.

[22] UNFCC website, above n 11.

[23] Boyd *et al*, above n 4.

resources, when too many donors have announced a setback in their level of aid disbursement.

Unfortunately, however, the use of aid to tackle climate change also raises some problems for developing countries. Firstly, only those countries that are ready to implement climate change projects will benefit from climate change financing, including aid (this so far has meant that richer countries such as China, India and Brazil have been allocated the biggest portion of climate funds[24]) this will not necessarily coincide with the countries that need the most aid. Secondly, ownership and partnership of aid might be compromised[25]—there might be an incentive to prefer projects that attract financing according to donors' interests (and, possibly, legal obligations) to tackle climate change rather than developing countries' real needs and development plans. This could be so especially if granting climate aid will be linked to donors' legal rather than moral responsibilities. Finally, as denounced by NGOs such as Oxfam the lack of developing countries' participation in the governance of the adaptation funds means that mitigation and adaptation projects are 'not demand-led, but instead driven by donor's priorities and preferences',[26] which results in a lack of developing countries' ownership of the project financed.[27]

WHAT AID INSTRUMENTS FOR CLIMATE CHANGE?

Since 1998, the DAC of the OECD has monitored aid targeting climate change and the Rio Conventions through its Creditor Reporting System (CRS)—its statistical system. The DAC/OECD has distinguished climate aid from development aid by looking at the main purpose, *raison d'etre*, of the aid granted. Using the Rio Markers, the OECD has gathered the different categories of aid as (i) targeting the Conventions as a 'principal objective' or a 'significant objective', or (ii) not targeting the objective.[28] The OECD individuates the first category of aid by asking donors whether the aid would have been disbursed were it not for the [climate change] objective.[29]

[24] A Ghosh, 'Harnessing the power shift: governance options for international climate financing' (6 October 2010) Oxfam Research Reports, available at www.oxfam.org.uk/resources/policy/climate_change/downloads/harnessing-the-power-shift-climate-finance-061010.pdf.

[25] Oxfam, above n 17.

[26] Oxfam, above n 17.

[27] Oxfam, above n 17. For those familiar with the aid effectiveness debate this goes against the principles endorsed in Paris and Accra.

[28] OECD, 'Measuring Aid Targeting the Objectives of the Rio Conventions' (May 2009) available at http://www.oecd.org/dataoecd/46/13/42819225.pdf.

[29] OECD, ibid 1.

international practice in this respect. Also the OECD, in its latest statistics on climate aid, seems to have adopted the notion of aid as public finances with a minimum grant element of 80 per cent, distinguishing the various aid instruments used simply by looking at the scope (intended as purpose) of the aid donated. It is submitted that such an approach, with the limitations and the problems highlighted already and further discussed below, is to be accepted and it will be used here for our taxonomy.

Climate Aid: Mitigation and Adaptation Aid

As highlighted by the literature climate aid is aid allocated to deal with specific environmental issues created by climate change.[35] Climate aid is the purest form of aid for climate change, where the primary objective of the aid granted is to address the negative effects of climate change. In this category one should include aid granted for funding both adaptation and mitigation projects.[36]

For example, a financial contribution by a government body in the form of a grant (or a loan with a grant element of at least 80 per cent of concessionality level) disbursed to a developing country to reduce greenhouse carbon gas emissions by setting up projects for alternative energy will fall into the category of climate aid. In order to classify this type of financing as aid, it is first important that the OECD definition of aid as regards its financial component is met, ie the financial contribution needs to be public money with a grant element of at least 80 per cent; it must be granted by a government body and it must be given to a developing country. Then the objective, the reason why the aid is donated, is the element that distinguishes climate aid from other forms of aid.

Mitigation projects are aimed at reducing climate change by setting policies and strategies intended to reduce greenhouse gas (GHG) emissions and achieve greater energy efficiency (IPCC, 2007). Adaptation projects are instead aimed at facilitating the development of techniques that 'can help increase resilience to the impacts of climate change.'[37] An example of adaptation projects is the development of an 'ancient canalization technique' used in Peru that will help farmers both through drought and heavy rains.

However, adaptation aid has not always been included in the category of climate aid. For example, for the past 13 years the OECD has defined

[35] See above.

[36] As already mentioned above mitigation and adaptation projects are not financed exclusively by aid money. Aid funds are frequently used to finance a specific aspect of a mitigation or adaptation projects—for example the capacity building aspect of the project. Aid can also be used to integrate other types of financing (mitigation projects are usually financed through public and private funding, including revenues from CDM).

[37] UNFCC Technical Paper, above n 6.

climate aid as 'activities that contribute to the objective of stabilisation of greenhouse gas (GHG) concentrations in the atmosphere at a level that would prevent dangerous anthropogenic interference with the climate system by promoting efforts to reduce or limit GHG emissions or to enhance GHG sequestration. This marker therefore only relates to *mitigation* aspects.'[38] (emphasis added)

The OECD was not the only organisation to exclude adaptation aid from climate aid. Many practitioners also classify adaptation aid as a special form of development aid for climate change, which should be included in the category of mainstreamed development aid for climate change (the second category in our taxonomy).

This restrictive definition of climate aid is no longer widely accepted and adaptation projects are now included in the category of climate aid by the vast majority of scholars. This view could now also be endorsed by the OECD,[39] which, in early 2011, announced it would collect data for adaptation aid. However, the OECD has not specified whether the data collected will be classified as climate aid or if they will fall within the category of mainstream development aid for climate change. The fact that the OECD will collect data for adaptation aid is a first very welcome change in the OECD approach.[40] Data for adaptation aid will not, however, be available before the end of 2011 and therefore one cannot yet be sure in which aid category adaptation aid will fall for the OECD.

Although the links between development and adaptation interventions are very close, it is submitted that a separation between the two categories of aid is still preferable and adaptation aid should be included in the category of climate aid.

Allowing a broader definition of climate aid, which includes adaptation projects, would not only adhere to the letter of the Convention and the Protocol, but would also be wise on a practical level. Indeed, excluding adaptation aid from the category of climate aid might give rise to some problems. For example, pure climate change projects would be very few, resulting in donors not meeting their climate aid targets under the UNFCCC and the Kyoto Protocol. As a consequence, donors might be more reluctant to release aid data on climate aid with the consequence that they would be less transparent in disclosing information about their aid funding. For example donors might decide to include all forms of aid—even aid that has development as its main objective—into one big category—without distinction between development and climate aid just to avoid disclosing the real

[38] OECD, above n 28.

[39] When this paper was submitted at the Oñati conference in 2009 the OECD only collected data in relation to mitigation aid and limited its definition of climate aid to aid for mitigation projects.

[40] OECD, above n 28.

(low) level of aid donated for climate aid. A more relaxed approach to types of aid falling into the climate aid category might encourage donors to be more specific when reporting their aid disbursements, to adhere to the classification proposed, allowing monitoring and transparency of aid funding. Furthermore, considering adaptation projects as development aid risks also diverting development aid resources from development to climate change. As further specified below, this might result in donors diminishing the level of aid donated for development projects.

A further problem with the OECD definition of climate aid needs to be highlighted here. We have mentioned above that the presence in the OECD definition of climate aid of the 'significant objective' category (ie where climate change is a significant objective of the aid granted and not just the 'principal objective') weakens the strength and the reliability of the OECD definition. Including, in the climate aid category, aid funds which have climate change as a 'significant objective' rather than limiting the category of climate aid to funding which has climate change as its 'principle objective,' blurs the distinction between climate aid and development aid with mainstreamed climate change objectives. This will be even more so once the OECD starts collecting data for adaptation projects; if the OECD maintains the definition of climate aid based on the 'significant objective' criteria when classifying adaptation aid, there will be a greater risk of hampering the possibility to understand if the aid granted for climate change is additional to development aid resources or is granted in substitution of development aid, ie aid would have been granted anyway as development aid. It is submitted that the best approach would be to look only at the principal objective and if the principal objective of the aid funding is to address mitigation or adaptation problems then the aid should be classified as climate aid. The 'significant objective' criteria needs to be excluded as too blurred and vague, so that its inclusion would risk undermining the definition itself.

Adaptation to climate change has become an important issue only in the past few years; this might explain why the OECD has only recently included adaptation aid in its statistical reporting system. Adaptation has become a serious issue since the UN Nairobi Agreement in 2006. From that time, adaptation programmes have made considerable progress. The Bali Action Plan, adopted in December 2007, 'identifies adaptation as one of the key building blocks required for a strengthened future response to climate change to enable the full, effective and sustained implementation of the Convention through long-term cooperative action, now, up to and beyond 2012.'[41]

Both the Kyoto Protocol and the Convention are designed to assist countries in mitigating and adapting to the adverse effects of climate change.

[41] See UNFCC Technical paper, above n 6.

Funding is provided through the financial mechanisms of the Convention, operated by the Global Environment Facility (GEF)[42] and the Adaptation Fund Board (AFB).[43] Funding opportunities include:

— the GEF Trust Fund;
— the Least Developed Countries Fund (LDCF) under the Convention;
— the Special Climate Change Fund (SCCF) under the Convention;
— the Adaptation Fund (AF) under the Kyoto Protocol and managed by the AFB; and
— obviously there is bilateral cooperation

More than 12 international adaptation and mitigation funds have been established, governed by multilateral and bilateral organisations.[44] The variety of funding mechanisms through which adaptation and mitigation projects are financed is harshly criticised by NGOs and practitioners. Oxfam, for example, argues that 'Adaptation finance is currently delivered through a spaghetti-bowl of different bilateral and multilateral channels. Multilateral funds are administered by the Global Environment Facility (GEF) or the World Bank. The exception to this is the Adaptation Fund under the authority of the Conference of the Parties (COP) of the UNFCCC.'[45] This lack of harmonisation in aid financing increases the transaction and administrative costs of the funded project.

Development Aid for Climate Change

Development aid—both in the form of multilateral and bilateral aid—has been an important, albeit controversial, tool for development and for eradicating poverty.[46] The strong links between development and climate change render difficult any clear-cut separation between development and climate change projects. Some aid projects will have both a climate change and a development impact. Thus a project aimed at enhancing development can, at the same time, have a positive impact on climate change. For example, the creation of a floating garden resilient to flooding will address both developing countries' development targets and will help them adapt to

[42] See Global Environment Facility website available at www.thegef.org/gef/, accessed 8/5/2011.

[43] See UNFCC website at unfccc.int/cooperation_and_support/financial_mechanism/adaptation_fund/items/4264.php.

[44] MC Lemos and E Boyd, 'The Politics of Adaptation' in M Boykoff (ed), *Politics of Climate Change: A Survey* (Cambridge, Cambridge University Press, 2009) ch 6.

[45] Oxfam, 'Beyond Aid: Ensuring Adaptation to Climate Change Works for the Poor' (16 September 2009) Oxfam Briefing Paper, available at www.oxfam.org.uk/resources/policy/climate_change/downloads/bp_132_beyond_aid_150909_web.pdf.

[46] International Conference on Financing for Development (Monterrey, Mexico, 18–22 March 2002). W Easterly, *The White Man's Burden. Why the West's Efforts to Aid the Rest Have Done So Much Ill and So Little Good* (New York, Penguin Press, 2006).

climate change. How to distinguish between the two categories? One way could be to look at the stated prevailing objective and at the environmental impact of the aid project.

A new trend seems to be developing for applying or 'mainstreaming' climate change into the development agenda[47] and a new category of development aid for climate change, usually referred to as 'mainstreaming climate change in development aid,' has been growing in prominence in recent years.

A considerable literature is developing, focusing on how to mainstream climate change in development aid policies and on how to ensure that development aid projects have a positive impact on climate change. However, development and climate change targets are not always coincident. They can sometimes be in conflict. For example genetically modified foods (GM foods or GMO foods) crops could be a means to address food security but they could harm biodiversity.[48]

The main difference between development and climate aid projects lies in the objective pursued. While for development aid the main objective is to eradicate poverty and meet development goals, as we have seen above, the primary aim of climate aid is to address climate change through mitigation and adaptation projects. This second category of development aid for climate change, while maintaining as its primary objective the promotion of development goals, also takes into account climate change issues as secondary objectives by ensuring that the development project financed has a positive impact on climate change. The projects are not, however, primarily directed at addressing climate change.

This can be very borderline with the type of aid highlighted above at 4.2, especially in relation to adaptation aid, yet this is a different type of aid and the difference should be taken into account especially when considering how much aid is donated by each donor and when verifying whether donors are meeting their commitments under the UNFCCC and the Kyoto protocol or the 0.7 per cent UN target for development aid.

Although a clear-cut separation between climate change and development issues is hard to draw and the two objectives can simultaneously be achieved, insisting on the difference and the separation of the two forms of aid is important to ensure that donors do not reduce their aid disbursements and, as pointed out by many, to ensure that the primary focus of development aid remains promoting development and the money spent on development are not reduced.

Many caution against the use of development assistance to fund climate change projects, because such a usage might 'divert funding away from other important development imperatives.'[49]

[47] Boyd *et al*, above n 4.
[48] I am grateful to Mark Halsey for raising this comment during the Oñati conference.
[49] Boyd *et al*, above n 4.

Enhanced transparency and clarity in the classification of aid granted could help better monitoring of aid activities and ensure that aid funds are not diverted from development activities.

Only by insisting on aid transparency requiring donors to state clearly the objectives of their aid projects and will it be possible to verify whether the commitments undertaken by donors are met. By excluding adaptation aid from the category of mainstream climate change in development aid and by including it in the climate aid category, we allow donors a broad leeway to meet their UNFCCC targets but at the same time by scrapping the OECD category of 'significant objective' from the climate aid category we eliminate ambiguities, avoiding confusion and diversion of funding. We also allow more transparency for development aid. In so doing we can also verify whether funding for climate aid is truly additional.

Climate Change As a Corollary Policy of Development Aid Project: Aid Conditionality and Climate Change Policies in the Procurement Process

Finally, a third category can be individuated in two different ways of combining development aid and climate change: namely, firstly in the insertion of climate change policies as a *conditio sine qua non* for the granting of development aid money and, secondly, in the use of climate change policies as secondary policies in the aid procurement process.

In particular I refer here to the use of aid conditionality (the former), where aid is granted on condition that certain climate change policies/standards are implemented, and to the use of the public procurement process to encourage the implementation of environmental/climate change policies (the latter).

The first category, namely aid conditionality, is well known in the development aid circle. This is linked to the use of aid to encourage developing countries' implementation of donors' preferred economic, social and human rights policies, where implementing the condition attached to the aid is a *conditio sine qua non* for the granting of the aid money, ie the aid will not be granted until and unless the recipient country accepts the policy set by the donor. Aid conditionality has a long and varied history, which goes from the Wold Bank's conditions to implement its (controversial and much criticised) structural reforms programmes to human rights conditionality.

Similarly, the disbursement of development aid money (hence aid that has development and not climate change as its primarily objective) could be subordinated to the implementation of climate change or environmental policies. While the use of aid conditionality is widely practised both formally and informally by donors, development experts are still divided about its effectiveness and its legitimacy. Many believe that imposing policy conditions on the granting of aid is an illegitimate interference with developing countries' right to self-determination and sovereignty. Moreover, many policies to which aid conditionality has been linked have proved to be unsuccessful—if not damaging—for developing countries (this has especially been so in regards to World Bank's policies). Others instead believe (especially human rights activists as regards the implementation of human rights policies) that the implementation of any such policies is essential for the success and correct implementation of certain economic and human rights instruments.

Making the granting of development aid money conditional upon the implementation of environmental and climate change policies risks raising the same debate. Have donors the right to impose on third countries their preferred environmental/climate change objectives especially given that they have spent the past 200 years polluting the world? Would such policies be really effective or would they just undermine recipient countries' ownership of aid? Would such use of development aid be compatible with the international environmental and climate change law instruments?

It is still too early to draw any final answer to the above questions; the use of development aid to foster environmental and climate change policies by attaching conditions to the aid granted is not widely diffuse. However it will be very important to carefully monitor this phenomenon if it were to expand. In this respect, the experience acquired from more than 50 years of implementation of aid conditionality in the development aid sector will surely prove useful.

The second category mentioned, namely the use of climate change policies in the aid procurement process, is related to making environmental and climate change standards part of the development aid procurement process, ie when spending aid money—for example, requiring that a certain percentage or a whole lot of aid contracts are reserved for enterprises that meet environmental/climate change standards, or giving price preferences to suppliers who meet environmental/climate change standards.

The use of the so-called horizontal/secondary policies[50] in public procurement is quite common. As Arrowsmith explains, many governments make use of public procurement as a policy tool to foster economic, social and environmental goals.[51] While the insertion of environmental objectives in the procurement process for general public procurement, ie procurement not related to aid money, is now an established phenomenon, the insertion of environmental/climate change policies in the aid procurement

[50] Horizontal/secondary policies are policy objectives other than the objective of the acquisition of goods/services on the best possible terms; see S Arrowsmith, 'Public Procurement as an Instrument of Policy and the Impact of Market Liberalisation' (1995) 111 *Law Quarterly Review* 235; for example, a national government may pursue social or environmental objectives through its public procurement.

[51] Arrowsmith, ibid.

process is a recent, and yet to be explored, phenomenon, which seems to be expanding considerably lately. Indeed, a real trend in development aid contracts seems to be developing for allowing (or advocating) the inclusion of such policies in aid procurement contracts. Although we are not yet at a point where these criteria are part of the award stage, they are already being considered by international soft law instruments. For example in the OECD Recommendation for untying aid, the updated 2008 version says at article 16 'those responsible for procurement should promote respect from suppliers with agreed international standards of corporate social responsibility and environmental behaviour. This could be done through reference to environmental and social considerations in tendering procedures.'

Many NGOs also call for the inclusion of environmental criteria in the procurement process. However, if such policies were to be included in the aid procurement process, would developing countries' industries be ready to implement such policies and to compete on equal grounds, with donors' enterprises and win aid contracts? There is a real risk that including such policies in the aid procurement process could favour donors' industries, because donors' industries are more familiar with such criteria and are more used to their own countries' (ie the donor) environmental standards. As such, they would be able to take advantage of the preference granted in the procurement process. Hence, including these requirements could have possible negative, protectionist, effects which would go against the principles and the spirit of the UNFCCC (see Art 3(5)). Making such criteria part of the aid procurement process, should be developing countries' exclusive choice, ie only the recipient of the aid should decide whether to insert price preferences or set aside contracts linked to environmental/climate change criteria, because only the recipient of the aid is in a position to know whether enterprises in its country would be able to take advantage of these conditions on an equal footing with donors' industries.

CONCLUSION

The Copenhagen accord calls for enhanced action and international cooperation on adaptation and mitigation projects. Donors are asked to increase their efforts to tackle the negative effects of climate change. However, as Boyd *et al* report, 'the funding committed so far is likely to fall far short of what is needed.'[52]

NGOs are adamant that funding for climate change should be additional to development aid funding, and not a substitute for it, but at times of constrained budget resources, and when many donors are cutting back on

[52] Boyd *et al*, above n 6.

development aid, achieving this target could be very difficult. NGOs and international organisations have no means to compel donors to fulfil their obligations (given that there is no coercive mechanism to insist that a country gives aid—it suffices to note that the UN target of 0.7 per cent GDP as a percentage of aid granted by countries is fulfilled only by a handful of donors); nor have they powerful instruments to monitor whether donors' commitments are met. The lack of clear and widely accepted classifications for aid devolved to climate change exacerbates this task. It is hoped that enhancing transparency when setting the objectives of aid projects can help to identify whether donors are disbursing aid for climate change and are fulfilling their promises of increasing aid for climate change while not distorting financing from other forms of aid projects.

9

Climate Change: Effects on Mobility of EU Workers and the Need to Safeguard Supplementary (Occupational) Pension Rights

KONSTANTINA KALOGEROPOULOU*

INTRODUCTION

CLIMATE CHANGE IS one of the most talked about challenges that humanity faces today. Even amongst critics, there appears to be overwhelming evidence that climate change is occurring.[1] The effects of climate change will be evident in Europe and impact upon the European Union's (EU) Member States and EU citizens.

This paper highlights the links between climate change effects and EU environmental policies, to EU employment goals and objectives, and focuses particularly on mobility of EU workers.[2] In this context, a particular issue that arguably impedes the right to free movement of EU workers, that of cross-border portability of supplementary pension rights, is discussed.

Part I examines ways in which climate change policies and action taken to respond to environmental concerns and adapt to new emerging environmental circumstances may impact upon EU employment strategies and the overall aims and aspirations of the EU. It looks at the ways in which

* I would like to thank the organisers of the conference, Professor Stephen Farrall, Professor Duncan French and Dr Tawhida Ahmed, for the opportunity to participate in it, and Dr Tawhida Ahmed and Professor Matthew Humphreys for specific comments on earlier drafts. All mistakes are my own.

[1] N Stern (Stern Report), *The Economics of Climate Change* (London, HM Treasury, 2006); Intergovernmental Panel on Climate Change (IPCC) Fourth Assessment Report: Climate Change 2007.

[2] Overall for this article, the term 'worker' includes both workers, within the definition of a worker given by the Court of Justice, as a person who performs services, for a certain period of time, for and under the direction of another person in return for which he receives remuneration (Case C-66/85, *Lawrie-Blum* [1986] ECR 2121), as well as the self-employed people who move to another Member State for employment purposes under Art 49 TFEU.

climate change may affect employment and jobs, and how these may enhance mobility of EU workers. Mitigation and adaptation policies may increase occupational and geographical mobility of EU workers, as a consequence of their impact on the employment sector that could result in loss of certain employment posts and creation of new workplaces. In addition, adverse climate and geographical changes, may lead to migration of EU workers and citizens.

Part II introduces a particular issue that needs to be addressed in order to promote the EU goals discussed in this chapter, and which may become more important in the light of the possible mobility of workers across EU Member States due to climate change effects and policies. That is the issue of facilitating 'portability' of supplementary (occupational) pension rights of persons who move across different Member States to take up employment throughout their careers. It emphasises the need to create a legal framework on cross-border portability of relevant pension benefits towards promoting workers' mobility within the EU, while enhancing social protection. This would facilitate the right to freedom of movement and enable EU workers to respond to climate change implications, while promoting the EU's objectives on employment and sustainable development.

Part I: Effects on Employment and Mobility of EU Workers

Climate change presents a major challenge, with significant implications for Europe and the European Union. It is argued that despite measures taken to deal with climate change, some effects such as rising temperatures, adverse weather conditions and rising of sea levels are already apparent and may be irreversible.[3] While the severity of the effects of climate change may vary by region, specific areas can be identified as the most vulnerable.[4]

These effects will have implications not only on the natural environment but also on sections of the economy and society within and across European states. Depending on when and how action at European, national, regional

[3] Commission's Communication, 'Green Paper, Adapting to Climate Change in Europe—Options for EU Action' COM (2007) 354 final, 3; 'White Paper, Adapting to Climate Change: Towards a European Framework for Action' COM (2009) 147 final; Study by the European Trade Union Confederation (ETUC) and the Social Development Agency (SDA) *et al*, 'Climate Change and Employment—Impact on Employment of Climate Change and CO_2 Emission Reduction Measures in the EU-25 to 2030 SYNTHESIS' (2007).

[4] These include southern Europe and the Mediterranean Basin, where there are increases in temperature and reduced precipitation together with water scarcity; mountain areas, particularly the Alps, where rapid rises in temperature are leading to melting of snow and ice and changed river flows; and coastal zones, as the sea levels rise and there are an increasing number of storms. Scandinavia, where more precipitation is expected and the Arctic region, where the highest changes of temperature are evident are other such areas that may be particularly affected from the impacts of climate change. COM (2007) 354 final, 5, above n 3.

and local level is taken to respond to these challenges,[5] and on the extent of climate change that takes place, the economic and social costs may vary. Nevertheless, it is clear that the effects of climate change will impact upon individual citizens as well as business, industry and the services sector.[6]

The EU, has, within the years of its evolution, incorporated environmental concerns in its aims and objectives and has adopted specific policies towards achieving EU and internationally agreed goals and dealing with climate change effects.[7]

Such environmental action and policies cannot be considered in isolation from the overall aspirations of the EU. Furthermore, climate change and the resulting effects pose new risks. These may underline further existing challenges the EU is facing, stemming from the recent financial and economic crisis that has led to a further increase in unemployment rates, the challenge of ensuring sustainability of public finances, the need to accommodate an ageing population, and the need to combat poverty.[8]

Since 1997, with the Treaty of Amsterdam, the EU has included sustainable development as an overarching objective of EU policies.[9] In order to achieve sustainable development,[10] economic development, social cohesion and environmental protection policies must mutually reinforce each other.[11] This is underpinned by the integration principle contained in Article 11 TFEU,[12] according to which, 'environmental protection requirements must be integrated into the definition and implementation of the Union policies and activities, in particular with a view to promoting sustainable development.'[13]

[5] COM (2007) 354 final and COM (2009)147 final, above n 3.

[6] COM (2007) 354 final, above n 3, 10.

[7] See generally, ec.europa.eu/dgs/clima/mission/index_en.htm; *Europe on the move* European Commission, Directorate-General for Communication, *Combating climate change, the EU leads the way* (2007).

[8] See for example Commission's Communication, Europe, '2020 A Strategy for Smart, Sustainable and Inclusive Growth' COM (2010) 2020 final, that identifies these issues.

[9] Art 3(3) TEU. In 2001 the EU adopted an EU sustainable development strategy, see Commission's proposal to the Gothenburg European Council, *A Sustainable Europe for a Better World: A* European Union Strategy for Sustainable Development COM (2001) 264 final; See also Commission's Communication, 'Mainstreaming Sustainable Development Into EU Policies: 2009 Review of the European Union Strategy for Sustainable Development' COM (2009) 400 final.

[10] Defined as 'development that meets the needs of the present without compromising the ability of future generations to meet their own needs', World Commission on Environment and Development (Brundtland Report, 1987).

[11] As a general overview of the mutually reinforcing pillars of sustainable development see M Lee, *EU Environmental Law, Challenges, Change and Decision-Making* (Oxford, Hart Publishing, 2005).

[12] Originally introduced with the Treaty of Maastricht in the Environmental Policy Title.

[13] The strategy for integration of environmental concerns COM (1998) 333 into EU policies was launched in Cardiff. The objective was introduced initially in the Single European Act and was given priority in the 5th Environmental Action Programme. On the integration of sustainable development into EU policy-making, see A Liberatore, 'The integration of sustainable development into EU policy-making' in *The Politics of Sustainable Development, Theory, Policy and Practice within the European Union*, S Baker, M Kousis, D Richardson

This section highlights the links between climate change and EU environmental and employment policies. It further focuses on the impact these may have on migrant workers who move within the EU for employment purposes.

Similarly to Article 11, Article 147(2) TFEU, contained in the Employment Title,[14] emphasises that 'the objective of a high level of employment shall be taken into consideration in the formulation and implementation of Union policies and activities.' EU environmental policies and measures to confront climate change, can therefore affect and help promote employment policies evolving around Article 145 TFEU, requiring Member States and the Union to develop a coordinated strategy for employment and to work towards promoting a skilled, trained and adaptable workforce and labour markets responsive to economic change, in order to realise the EU objectives. In addition, according to Article 9 TFEU, the Union needs to take into account requirements linked, amongst other goals,[15] to the promotion of high levels of employment and the guarantee of adequate social protection, when defining and implementing its policies and activities. These aims further interrelate with one of the ultimate goals of the European Union, the right of free movement of workers.

The EU, in its new strategy Europe 2020,[16] included, within the context of promoting inclusive growth as one of three suggested mutually reinforcing priorities, the aim of achieving a high-employment economy that would enhance social and territorial cohesion.

In 'an economy based on knowledge and innovation', and sustainable growth, focusing on 'promoting a more resource efficient, greener and more competitive economy' that includes stimulating the creation of green jobs, inclusive growth will help to enhance the EU's 'potential for sustainable growth and competitiveness'.[17]

In order to 'catalyse progress', a new tool was introduced in the form of flagship initiatives.[18] One of the seven flagship initiatives entails an agenda for new skills and jobs, focusing on the aim of modernising labour markets in order to raise employment rates and ensure the sustainability of social

and S Young (eds) (Abingdon, Routledge, Global Environmental Change Series, 1997); On the history of environmental policy integration, see M Herodes, C Adelle and M Pallemaerts, *Environmental Policy Integration at the EU Level—A Literature Review*, EPIGOV Paper No 5 (London/Brussels, IEEP, 2007).

[14] Introduced in the Treaty of Amsterdam.

[15] These include the fight against social exclusion, and a high level of education, training and protection of human health.

[16] Above n 8.

[17] The Commission proposed five headline interrelated targets towards meeting the aims of the Europe 2020 strategy. See above n 8.

[18] See ec.europa.eu/europe2020/tools/flagship-initiatives/index_en.htm.

needs.[19] This would require the development of a skilled workforce and the promotion of productivity and employability that would enable people to meet the needs of modern labour markets.[20]

Modern labour markets require flexibility and security[21] in the market to ensure that all workers, employers and Member States benefit from employment opportunities and a high level of employment is achieved. Workers should be encouraged to improve their job prospects, while at the same time, employers and companies should be able to attract and recruit people that suit the demands of the market.[22] To achieve this, promoting labour mobility within the EU is essential to ensure that the EU workforce can adjust to changing labour markets. The demands of a global economy and the social trends and developments, including a rapidly ageing EU society,[23] require greater levels of mobility to promote employment and the adaptability of workers. Therefore, advancing occupational and geographical mobility and the improvement of existing legislation and administrative practices that impede such mobility is necessary for EU aspirations to be met.[24]

The modernisation of labour markets has to be seen in parallel with the modernisation of social security schemes that would enable workers to maintain and ameliorate their standard of living, by not being disadvantaged in relation to social security rights and entitlements (including pension benefits) and meet the aspirations of mobile workers.

Focus on growth and jobs, through the creation of more and better jobs, increasing adaptability of workers and the promotion of flexible labour markets including through workers' mobility and the modernisation of labour markets and social protection systems, have attracted particular focus at EU level, since the 1990s.[25] These goals were further reinforced with the Lisbon strategy on Europe to become 'the most competitive and

[19] To facilitate the functioning of Europe's labour markets the Commission has proposed concrete actions. See ec.europa.eu/social/main.jsp?langld=en&catld=958.

[20] See also Commission's Proposal for Guidelines for the Employment Policies of the Member States, *Part II* of the Europe 2020 Integrated Guidelines: COM (2010) 193 final.

[21] See Commission's Communication, 'Towards Common Principles of Flexicurity: More and Better Jobs Through Flexibility and Security': COM (2007) 359 final.

[22] COM (2007) 354 final above n 3.

[23] See Commission Staff Working Document, Demography Report 2010, *Older, more numerous and diverse Europeans*, available at www.europa.eu and Commission's Communication, *Dealing with the impact of an ageing population in the EU* (2009 Ageing Report), COM (2009) 180 final.

[24] See Commission's Communications, 'New Skills for New Jobs Anticipating and Matching Labour Market and Skills Needs' COM (2008) 868 final; 'An Agenda for New Skills and Jobs: A European Contribution Towards Full Employment' COM (2010) 682 final.

[25] Especially with the Treaty of Amsterdam and the inclusion of a new Title on Employment Title (and the adoption of the European Employment Strategy) and the incorporation of the Social Policy Agreement. See also Commission's Communications, 'European Social Policy, A Way Forward for the Union, A White Paper' COM (94) 333 and Concerted strategy for modernizing social protection: COM (1999) 347 final. On the European Employment Strategy see ec.europa.eu/social/main.jsp?catId=101&langId=en.

dynamic knowledge based economy in the world, capable of sustainable economic growth with more and better jobs and greater social cohesion'[26] and the renewed Lisbon goals, that centred on growth and more and better jobs.[27] The realisation of these goals also required an open and accessible labour market,[28] and a skilled and adaptable labour force[29] that could meet its demands.

To this end, barriers to occupational and geographical mobility of workers need to be addressed in order to promote the mobility of workers within the EU. Such obstacles derive from the inability to safeguard supplementary pension rights for migrant workers that will be discussed in Part II.

Climate change effects and EU environmental policies may both affect and impact upon EU goals relating to the promotion of employment and inclusive growth identified above, and emphasise the need to enhance workers' mobility.

Potential effects of climate change may create new circumstances for EU citizens and workers. A number of economic sectors have been identified as likely to be affected by climate change. These include agriculture, forestry, fisheries and aquaculture sectors, human, animal and plant health, and the energy sector. In addition, tourism, insurance, finance and infrastructure (including buildings, transport and industrial infrastructure and energy and water supply) are also likely to be affected.[30] The effects on these sectors and on relevant economic activity will also be evident in employment.[31] Even gradual and moderate climate change may impact on employment, with both negative and positive consequences.[32]

The potential consequences of climate change and environmental policies for employment have been underlined on various occasions.[33]

[26] Presidency Conclusions, Lisbon European Council (23–24 March 2000).

[27] See Presidency Conclusions, above n 26 and Commission's Communication, 'Working Together for Growth and Jobs, A New Start for the Lisbon Strategy' COM (2005) 24.

[28] See for example Commission's Communications, 'New European Labour Markets, Open to All, With Access for All' COM (2001) 116 and 'Mobility, an Instrument for More and Better Jobs: the European Job Mobility Action Plan (2007–2010)' COM (2007) 773 final.

[29] See for instance Commission's Communication, 'Action Plan for Skills and Mobility' COM (2002) 72 final and COM (2007) 773 final, above n 28.

[30] COM (2007) 354 final and COM (2009) 147 final above n 3; study by ETUC and SDA, above n 3.

[31] See A Rosemberg, 'Building a Just Transition: The linkages between climate change and employment' (2010) 2 *International Journal of Labour Research* 125, 130; and Study by ETUC and SDA, above n 3.

[32] See study by ETUC and SDA, above n 3. See more generally, OECD, Working Party on National Environmental Policy, *Environment and Employment: an Assessment* (ENV/EPOC/WPNEP (2003) 11/FINAL 2004) www.oecd.org/dataoecd/13/44/31951962.pdf.

[33] Commission's Communication on 'Environment and Employment (Building a Sustainable Europe)' COM (97) 592 final; Commission Staff Working Document on the Links between employment policies and environmental policies: SEC (2005) 1530; Report by UNEP, ILO, IOE, ITUC, *Green Jobs Towards Decent Work in a Sustainable, Low-Carbon World* 2008; M Rayment, E Pirgmaier, G De Ceuster, F Hinterberger, O Kuik, H Leveson Gower, C Polzin and

In order to respond to climate change effects, mitigation and adaptation policies may present new opportunities for workers and self-employed persons within the EU. In addition, EU environmental policies may also promote and create more employment opportunities, through, for instance, the promotion of new environmental technologies, and innovative products and services, or environmental regulation that may affect businesses and create more working places.[34] These may lead to the creation and development of business and promotion of sustainable production. Energy production, new infrastructure necessary to protect the environment and respond to climate change effects (like for instance the coastal protection or transport), and green tourism, are other economic sectors that may impact on employment and result in the creation of new working posts. Moreover, green jobs are created in the context of both the need to reduce CO_2 emissions, and adapt to climate change effects.[35] Further opportunities are provided through increased demand for insurance products and services to protect against extreme weather events as well as through research and empirical studies undertaken with regard to environmental and climate change effects.[36]

Nevertheless, climate change effects and environmental policies may also have negative consequences and lead to loss of jobs. The possible rise in sea levels, for example, may affect agricultural activities and tourism.[37] The inability to adapt to new challenges and the needs posed by climate change effects may affect companies of particular sectors and this may in turn lead to job loss for their employees. Workers and (self-employed persons) with no particular skills or training required for posts relating to environmental-related jobs may also be affected. Some sectors may be sooner or more abruptly affected.[38] Environmental policies may also have negative effects on businesses,[39] and result in rising unemployment levels. Although suggestions focus on the short term and sectoral level, unexpected effects may diminish employment levels and prospects.[40]

A Varma, 'The Economic Benefits of Environmental Policy, a Project under the Framework Contract for Economic Analysis (Amsterdam, Institute for Environmental Studies, Final Report November 2009); GHK in Association With CE and IEEP, 'Links Between the Environment, Economy and Jobs' (Policy-makers Summary 2007) http://ec.europa.eu/environment/enveco/industry_employment/pdf/ghk_study_wider_links_summary.pdf.

[34] See SEC (2005) 1530, above n 33.

[35] See COM (97) 592 final, above n 33; Study by the ETUC and the SDA, above n 3; see also Commission's Factsheet 'What social Europe can do for you' (A6) More and Better Jobs-Green Jobs' (2010).

[36] See SEC (2005) 1530, above n 33.

[37] See study by ETUC and SDA, above n 3.

[38] Tourism is an area where the effects of climate change are already evident (COM (2007) 354 final and COM (2009) 147 final, above n 3).

[39] Through for instance environmental regulation, see SEC (2010) 1530, above n 33.

[40] OECD Working Party on National Environmental Policy, 'Environment and Employment: an Assessment' (2004); also other studies (see for example SEC (2005) 1530, above n 33, 10),

The possible effects on employment identified above reinstate the interrelation between climate change effects and policies adopted at EU level, and the attainment of EU employment goals, which focus on achieving the overarching EU objectives of reducing unemployment and increasing labour participation and productivity, while at the same time promoting sustainable development and addressing environmental concerns and aims.

The focus of environmental policies on environmentally-friendly technology, goods and services requires, for example, the preparation of qualified workers with the necessary skills and competencies to respond to emerging and future environment-related employment and 'green economy'[41] needs and the promotion of quality jobs.[42] Such trends may also impact on goals of promoting investment in lifelong learning. Improving education and training systems is necessary in order to embrace and encourage the EU workforce, while reducing inequalities,[43] and allows for both workers with limited skills and experienced high-qualified people to adapt to new developments put forward by environmental changes and demand.

This chapter focuses particularly on the mobility of workers across EU Member States. Geographical and occupational mobility within the EU may be underpinned as a result of EU policies and responses to environmental concerns and climate change effects described above, and new circumstances that these policies and responses may give rise to.

EU workers may have to move, within and between economic sectors, in order to take up new employment, either because of loss of previous employment or due to better and more appealing job prospects. Therefore, the need to be flexible and to adapt in order to benefit from new employment opportunities presented as a result of environmental policies and climate change effects may become more pressing. In this context, the protection of social security rights of migrant workers is necessary to enable them to move between Member States, without diminishing their level of social protection.

Mobility may also be a result of environmental changes and adverse climate change effects (such as the erosion of coastal lines or temperature thresholds being exceeded, making living in those areas difficult or impossible),

indicate that the overall job impact might be minimal or not occur. Nevertheless, it may be argued that abrupt changes may have more severe impacts on employment.

[41] See Rosemberg, above n 31.

[42] See ECORYS A Slingenberg, K Rademaekers, E Sincer and R van der Aa, 'Environment and Labour Force Skills, Overview of the Links Between the Skills Profile of the Labour Force and Environmental Factors', final report (2008) http://ec.europa.eu/environment/enveco/industry_employment/pdf/labor_force.pdf.

[43] Commission's Communication, 'Efficiency and Equity in Education and Training Systems' COM (2006) 481 final.

particularly in densely populated areas.[44] Workers' migration, as a potential result of climate change, may be seen both as a way of adapting and conforming to emerging social, economic, cultural and environmental circumstances, as well as from a negative point of view, the failure to adapt to changes within the physical environment.[45] The number of mobile workers may also depend on mitigating action and the actual realisation of the predicted risk of climate change or level of intensity of climate change effects. More detrimental effects of climate change may have further unpredictable negative implications for these sectors, especially if adaptation policies are not introduced and followed. Nevertheless, even moderate impact may possibly result in workers moving to other regions and Member States for employment. Therefore, facilitating workers' mobility, to enable them to adjust and integrate in new environments and moderate the negative consequences in relation to employment prospects and rights, acquires again particular importance.

To this end, it can be argued that enhancing labour mobility by removing existing barriers to the exercise of the right to free movement—which constitutes one of the main goals of the EU since the inception of the European Economic Community—and rewarding professional and geographical mobility, become more important in order to address current and future needs created by environmental and climate change developments and risks in the labour market. This would enable workers to adapt to new circumstances or take advantage of new employment opportunities, and help to achieve the Union's objective of full employment, by increasing labour participation and responding to the market needs of demand and supply, allowing it to attract a trained and capable workforce, while securing social protection of EU citizens and helping to promote social cohesion. Such obstacles to labour mobility arise due to the absence of a legal framework enhancing cross-border portability of supplementary pension rights for EU workers, which this chapter will focus on in the next section.

Part II. Meeting the EU Goals: Workers' Mobility and the Need to Safeguard Supplementary Pension Rights

This section highlights the need for a legal framework that would enhance cross-border portability of supplementary pension rights and could consequently facilitate the right of freedom of movement and occupational

44 COM (2007) 354 final and COM (2009) 147 final above n 3; study by ETUC and SDA, above n 3.

45 C Tacoli, 'Crisis or Adaptation? Migration and Climate Change in a Context of High Mobility', Prepared for Expert Group Meeting on Population Dynamics and Climate Change UNFPA and IIED, in *Collaboration with UN-HABITAT and the Population Division, UN/DESA* (24–25 June 2009) www.unfpa.org/webdav/site/global/users/schensul/public/CCPD/papers/Tacoli%20Paper.pdf.

and geographical mobility of workers. It considers how the EU goals discussed above and the potential implications of climate change effects and policies on employment—whether temporary or permanent, voluntary or compulsory—which fortify the need to promote existing EU policies on enabling mobility of EU workers, underpin the importance of promoting cross-border portability of supplementary pension rights. It further argues that facilitating portability of such pension entitlements may contribute to the response to the implications of climate change effects for mobility of workers, identified above.

Within the context of this chapter, the term 'supplementary pensions' refers to pension benefits related to the workers' employment relationship or the exercise of a profession. They aim to supplement (or partly substitute, depending on the level of coverage) state-provided pension income and are described as occupational pensions.[46]

Cross-border portability of supplementary (occupational) pensions can be linked to the right of freedom of movement of EU citizens across EU Member States, one of the fundamental goals of the EU. It encompasses the possibility for migrant workers, who move across Member States, to acquire and preserve their supplementary pension benefits.[47]

[46] This classification follows the focus of various documents of the European Commission that attempted to draw attention to cross-border portability of supplementary pension rights. See for example Commission's Communications: 'Supplementary Social Security Schemes: the Role of Occupational Pension Schemes in the Social Protection of Workers and Their Implications for Freedom of Movement' SEC (91) 1332 final; Commission's Green Paper, Supplementary Pensions in the Single European Market: COM (97) 283; 'Results of the Consultations on the Green Paper on Supplementary Pensions in the Single Market' COM (1999) 134 final; 'Proposal for a Directive on Improving the Portability of Supplementary Pension Rights' COM (2005) 507; Amended proposal for a directive on minimum requirements for enhancing worker mobility by improving the acquisition and preservation of supplementary pension rights: COM (2007) 603. These types of pension are broadly classified (in the Commission's documents categorization) as pillar two within the EU Member States' national structures on retirement systems; pillar one concerns mainly statutory pensions, while the third pillar also concerns supplementary private pensions that have been individually arranged. The three pillars may be highly interdependent and distinctions between pension schemes may be difficult to make. See Report by the Social Protection Committee (SPC), on 'Privately Managed Pension Provision' (2005) and study on Privately Managed Funded Pension Provision and their Contribution to Adequate and Sustainable Pensions (2008). Other classifications into three or more categories exist (see classifications by the World Bank, ILO, OECD).

[47] For an introduction to the topic see V Andrietti, 'Portability of Supplementary Pension Rights in the European Union' (2001) 54 *International Social Security Review* 59; J Jolliffe, 'The Portability of Occupational Pensions Within the European Community' (1992) *Transactions of the International Congress of Actuaries* 131; CEPS Task Force Report, 'Cross-border Portability of Pension Rights, An important Condition for an Integrated Market for Pension Provision', Rapporteur: J Mortensen, (no 45, 2003); J Marshall and S Butterworth, 'Pensions Reform in the EU: The Unexploded Time Bomb in the Single Market' (2000) 37 *Common Market Law Review* 739. Portability of supplementary occupational pension rights may also refer to intra-state portability (see COM (2005) 507 above n 46). The chapter does not, however, focus on this type of occupational mobility.

Currently, workers who take up employment in different Member States throughout their careers may be disadvantaged in the supplementary pension benefits that they will receive once they reach retirement age, compared to workers that remain with the same employer and in the same Member State.[48] It can therefore be argued that the lack of a legal framework safeguarding supplementary pension benefits that may lead to reduced and fragmented pension entitlements, impedes workers' mobility and discourages them from moving within the EU.[49]

The main obstacles to cross-border portability of supplementary (occupational) pension rights concern the acquisition, preservation and transferability of such rights. In many occupational pension schemes, certain qualifying conditions exist in order for an employee to be able to subscribe to the pension scheme.[50] Workers who move to a different State for employment and leave a particular employment relationship without having fulfilled these conditions will not have any vested rights in a pension income.[51] Obstacles to preservation of such rights derive from the lack of measures ensuring that acquired pension benefits that are 'left' in the last pension scheme retain their value.[52] Transferability refers to the possibility of a capital value that represents the pension entitlements of an employee, being transferred to another scheme or to a similar financial institution and is closely related to taxation practices that may pose further obstacles to portability of supplementary pension rights.[53] These hindrances can be attributed to the great heterogeneity that exists within supplementary pension schemes[54] and the different ways occupational schemes operate in the various Member States.[55]

[48] Statutory pension schemes (and second pillar pension schemes that operate on a pay-as-you-go basis) on the other hand, were covered by Regulation 1408/71, on co-ordination of national social security legislation (repealed by Regulation 883/2004/EC).

[49] COM (2005) 507, above n 46.

[50] These concern the waiting period and the minimum age (the period of time or age required before an employee can subscribe to an occupational pension scheme) demanded, and the vesting periods required for workers to be eligible for participation in such schemes. COM (2005) 507, above n 46.

[51] These practices aim to reward loyal employees but are also used to avoid excessive fragmentation of supplementary pension rights, and disproportionately burdensome administration for the funds.

[52] This impacts mainly on workers who change jobs but leave their vested rights in the scheme relating to the previous employment.

[53] See Commission's Communication, 'The Elimination of Tax Obstacles to the Cross-Border Provision of Occupational Pensions' COM (2001) 214 final.

[54] This is a result of their different development, according to social, political, economic and historical background. See H-D Steinmeyer, 'The Variety of Occupational Pension Systems in the EC Member States as an Obstacle for Mobility?' in M Brennan (ed), *1992 and Beyond: An Explosion in Employee Benefits?* (CEBS, 1990).

[55] See the Report and Study by the SPC, above n 46. The organisation of social security schemes lies within the competence of each Member State that decides by what combination of statutory and supplementary schemes the objectives of social protection are to be met, while

Further to the right to freedom of movement and market integration aims,[56] supplementary pensions and consequently supplementary pension rights, have attracted attention due to the goal of modernising social protection, which was reaffirmed with the launch of the Lisbon strategy,[57] and the focus on ensuring adequate, safe and sustainable pensions.[58]

On the one hand, pension systems need to evolve to respond to the changes in the labour market and the changes in the traditional concept of life-long careers in the same employment position. As modern labour markets require more flexibility and security in order to respond to the aspirations of mobile workers within the EU, pension systems should not penalise job mobility, as may be the case with regard to the supplementary pension benefits of migrant workers who are suffering losses in their rights. Removing obstacles to cross-border portability of occupational pension rights would provide an incentive for persons to move to another Member State, when taking into consideration the financial implications of such a decision for their pension income, and would consequently help to facilitate and improve geographical and occupational mobility.

On the other hand, societal changes also call for the modernisation of social security systems and of pension provision, in order to guarantee sustainable and adequate pensions, one of the main social challenges the EU is currently facing.[59] Supplementary pensions have attracted a lot of attention as a result of demographic trends that place the sustainability of national social security systems at risk. The increase in life expectancy and the decrease in fertility rates, coupled with an ageing working population, may hinder the advancement of employment and economic growth, while requiring further resources to provide for an elderly population.[60] This

the EU lacks direct competence to intervene in pension issues. Moreover, the significant role that social partners hold in some types of occupational pension schemes and the principle of subsidiarity (Art 5 (3) TEU), according to which, any action taken at EU level on this issue, should not go beyond what is necessary to achieve the intended objective, further restrict EU interventions.

[56] Portability of supplementary pension rights also relates to the freedom of provision of services, of establishment (as pension providers may be operating in a different Member State than the one where the employer is based) and of capital as a worker may need to transfer the whole capital he has invested in a previous scheme.

[57] Above n 26.

[58] See for example Commission's Communications, 'The Future Evolution of Social Protection From a Long-Term Point of View: Safe and Sustainable Pensions' COM (2000) 622; 'Supporting National Strategies for Safe and Sustainable Pensions through an Integrated Approach' COM (2001) 362; See also 'Commission's Green Paper Towards Adequate, Sustainable and Safe European Pension Systems' COM (2010) 365 final and accompanying document: SEC (2010) 830.

[59] The need for sustainability of pension systems concerns both state-provided and supplementary pensions (see the study by the SPC (2008), above n 46).

[60] This is due to the fact that statutory first pillar pay-as-you go schemes depend on solidarity between the generations, as contributions by current workers are used to provide payments for those who have retired and consequently require a balanced analogy of workers

may have a significant impact on national budgets and the Member States' capacities to provide an adequate level of social protection. Supplementary pension provision is considered to shift some of the financial burden from the Member States that are in principle responsible for the state pensions, to funded supplementary pensions related to employment (occupational pensions).[61] This has led to the increased use of such schemes and eventually to a general global trend of reform of pension schemes towards promoting private (supplementary) schemes.[62]

It can be argued that potential climate change and policies' effects on employment and the importance of promoting and facilitating labour mobility analysed above, underpin the need for an effective legal framework on acquiring and preserving supplementary pension entitlements. This could reduce obstacles to the mobility of workers and enable them to both respond to environmental related impacts on the employment sector, taking advantage of new opportunities created and fulfilling the needs of the market, and to adjust to new circumstances created, while limiting the adverse effects on their standard of living and their future pension income.

In addition, ensuring safe and sustainable pensions, and in this way enhancing social cohesion, which also forms part of the EU goals relating to sustainable development, can become more challenging as a result of economic effects and implications of climate change.[63] Member States may have to bear a further economic burden, in order to prepare or confront climate change threats and effects.[64] Further proactive measures could

and entitled pensioners. In supplementary occupational pensions, on the other hand, the savings of active workers are invested and are used to finance the future pension income of those workers. The decline in the birth rate results in fewer workers contributing to the national budget, while the rise in life expectancy leads to pension benefits being paid for longer to retired people. This may lead to a significant increase in public spending on pensions for the Member States. The Stability and Growth Pact places further pressures on the Member States. See Directorate-General for Economic and Financial Affairs of the European Commission, 'The 2009 Ageing Report—Economic and Budgetary Projections for the EU-27 Member States (2008–2060) (2009) 2 European Economy; Commission's Communication, 'The Demographic Future of Europe—From Challenge to Opportunity' COM (2006) 571 final.

[61] Or to individual schemes (third-pillar schemes). See the Report and the Study by the SPC, above n 46.

[62] Y Stevens, G Gieselink and B Van Buggenhout, 'Towards a New Role for Occupational Pensions in Continental Europe: Elements and Techniques of Solidarity Used Within Funded Occupational Pension Schemes' (2002) 25 *European Journal of Social Security*. Private pension schemes can also be affected by demographic changes (see COM (2010) 365 final, above n 58, 4) and are vulnerable to financial risks (as the recent financial and economic crisis has demonstrated). Various steps have been taken to ensure appropriate supervision of financial institutions. See for example Directive 2010/78/EU, OJ 2010 L331/120.

[63] See Stern Report, above n 1; 'Facts and Figures: The Links between EU's Economy and the Environment' (2007) available at ec.europa.eu/environment/enveco/industry_employment/pdf/facts_and_figures.pdf.

[64] The effects of climate environmental changes may be more evident in poorer areas where there are fewer financial resources to develop policies and adopt measures to confront and adapt to new circumstances.

be required for more vulnerable areas.[65] The public sector may have to respond to arising difficulties,[66] which can be further augmented if co-ordinated action at EU level, for instance, does not take place or proves inadequate. Such economic implications could pose severe restraints on Member States' finances, with particular effects on the delivery of adequate welfare provision. The EU is already trying to deal with the implications of the recent economic and financial crisis that have increased debts and national economic deficits. This is exacerbated by the rise of unemployment at EU level, which could further increase, particularly in certain areas, as a result of the climate change impacts. In addition, enhanced migration, as a possible consequence of environmental changes and policies identified above, can result in additional pressures on state pension provision, which is based on inter-generation solidarity, especially where a large number of workers seek employment in other Member States.

In this context, cross-border portability of supplementary pension rights becomes again important, in order to ensure that supplementary pension benefits are not diminished and that workers who take up employment in different Member States will enjoy an adequate standard of living upon retirement. Safeguarding supplementary pension rights would allow workers to build up adequate pension income during their employment relationship with each employer, before environmental-related employment circumstances result in loss of current workplaces, and it would help to avoid or reduce fragmentation of supplementary pension rights, once they start to move within the Member States to take up new employment. Removing obstacles to the acquisition and preservation of supplementary pension entitlements would also make such pension provision more attractive for workers and help it develop and contribute towards achieving safe and sustainable pensions.

The need to facilitate cross-border portability of supplementary pension rights in order to enhance mobility of EU workers has been identified on various occasions within the context of achieving the EU goals of full employment and growth and jobs,[67] while improving social protection.[68] These goals, in conjunction with other EU objectives and policies, towards creating a sustainable future for Europe. Despite various attempts[69] to adopt legally binding measures dealing with specific obstacles to cross-border portability

[65] For example appropriate infrastructure in the areas on the coastal zone or transport infrastructure.

[66] COM (2007) 354 final, above n 3, 10.

[67] Above n 29 and COM (2005) 24, above n 27.

[68] See Commission's Communication, 'Social Policy Agenda' COM (2000) 379; Commission's Communication on the 'Social Agenda' COM (2005) 33 final; Commission's Communication, 'Renewed Social Agenda: Opportunities, Access and Solidarity in 21st Century Europe' COM (2008) 412 final.

[69] See above n 46.

identified above, it has proven a great challenge to achieve consensus,[70] and an adequate legal framework to enhance portability is not yet in place.[71] It can be argued that enhanced mobility across the EU and restraints on national economies and welfare provision, as a possible result of climate change effects, may further underpin the need for action to be taken towards facilitating such portability.

In October 2005 a long-awaited proposal on improving the portability of supplementary pension rights was put forward by the European Commission, dealing with some of the main obstacles identified above.[72] This could have constituted a first step towards the creation of a more effective legal framework on portability of such rights. A revised proposal was published in 2007, omitting the transferability element and changing the scope of the proposal from portability to 'enhancing worker mobility by improving the acquisition and preservation of supplementary pension rights'.[73] Agreement in Council has not been reached, even on the amended proposal, although the proposal still 'remains "live"'.[74] The need to deal with this issue has again been identified within the context of achieving the goals of Europe 2020.[75]

[70] E Oliver, 'From Portability to Acquisition and Preservation: The Challenge of Legislating in the Area of Supplementary Pensions' (2009) 31 *Journal of Social Welfare and Family Law* 173.

[71] Only Directive 98/49/EC OJ 1998, L209/46, on safeguarding supplementary pension rights of employed and self-employed persons moving within the European Union has been adopted, which provides only for limited portability. Directive 2003/41/EC OJ 2003, L235/10, as amended on the activities and supervision of institutions for occupational retirement provision (IORPs), could facilitate labour mobility for some categories of employees, since it allows for the set-up of pan-European pension funds by multinational companies. See P Dejmek, 'No Flying Start but a Bright Future for EU Directive 2003/41/EC on Occupational Pension Institutions' (2006) 17 *European Business Law Review* 138; and I Guardiancich, 'Pan-European Pension Funds: Current Situation and Future Prospects' (2011) 64 *International Social Security Review* 15. However, the scope of this Directive is restricted, as it does not apply to all types of occupational pension schemes.

[72] COM (2005) 507 final, above n 46 and Commission staff working document, annex to the proposal SEC (2005) 1293. The proposal dealt with the conditions on acquisition, preservation and transferability of supplementary pension rights, as well as information provided to workers on how the termination of employment would affect their supplementary pension rights and focused on both intra-state and inter-states mobility. The taxation issues were not addressed. For an analysis of the proposal see K Kalogeropoulou 'Improving the Portability of Supplementary Pension Rights' (2006) *Journal of Social Welfare and Family Law* 95.

[73] For an analysis of the proposal see Oliver above n 70.

[74] COM (2010) 365 final and SEC (2010) 830 final, above n 58.

[75] See Commission's Communications, 'Towards a Single Market Act, for a Highly Competitive Social Market Economy, 50 Proposals for Improving Our Work, Business and Exchanges With One Another' COM (2010) 608 final, proposal no 31; EU, 'Citizenship Report 2010 Dismantling the Obstacles to EU Citizens' Rights' COM (2010) 603 final; COM (2010) 682 final, above n 24.

CONCLUSION

This chapter has highlighted the interrelations between evolving climate change effects and EU policies, focusing on the ways these may impact on EU employment goals and mobility of workers, and it has identified a particular issue relating to mobility and the right of freedom of movement of workers that acquires more importance as a result of climate change implications.

In particular, the chapter has demonstrated that climate change is likely to result in shifts in employment between old and new jobs and new employment opportunities or circumstances for EU workers. This may have significant impact on the mobility of EU workers across the Member States and, combined with the economic restraints that environmental developments placed on public finances, relate to existing EU goals, towards achieving flexible labour markets and a high level of employment, while ensuring social protection for EU citizens.

These considerations underpin the need to promote an existing issue that has been attracting attention at EU level for the last 20 years, that of enhancing cross-border portability of supplementary pension rights for EU workers. The lack of such portability may, arguably, impede labour mobility and diminish pension income and levels of social protection for EU workers.

Building an adequate legal framework to guarantee supplementary pension rights for EU workers may help to respond to new employment and social circumstances that can be created because of climate change, while contributing towards the realisation of the EU goals on achieving smart, sustainable and inclusive growth. Although it has not yet been possible to adopt legislative measures dealing with particular obstacles to cross-border portability of supplementary pension rights, the issue is still considered at the EU agenda.

Given the difficulties that arise with regards the adoption of binding legislative measures, and the limited success of relevant proposals, soft law, as another type of policy instrument, could, with its non-binding form, provide for an alternative way for the EU to engage in this area and promote cross-border portability of supplementary pension rights through the convergence of national policies on the issues that impede portability.[76] Soft law has already been used to raise interest and draw attention to the need to enhance cross-border portability of supplementary pension rights.[77] The open method of coordination (OMC) that provided for another soft law tool could also be considered a suitable way to promote convergence

[76] See K Kalogeropoulou, 'European Governance After Lisbon and Portability of Supplementary Pension Rights' (2006) 2 *Journal of Contemporary European Research* 75.

[77] See for instance above n 46.

of national practices towards reducing obstacles to acquiring and preserving supplementary pension rights.[78] In addition, the flagship initiatives, as a new tool presented in the EU's strategy 'Europe 2020' towards achieving the targets set, which include the aims of modernising the labour market and promoting labour mobility, may provide for another opportunity to address the issue of facilitating cross-border portability of supplementary pension rights.

There can be little doubt that climate change is now occurring. Some of the possible impacts of climate change have, however, been discussed more than others. Within the context of acknowledged and discussed effects of climate change on EU goals and objectives, this chapter has pointed out an area that has rarely been considered in relation to climate change. Nevertheless, in order for the European Union to prepare and address climate change effectively, all potential impacts, and the ways to mitigate resulting outcomes, need to be addressed. The aim of achieving and promoting sustainable development requires a comprehensive strategy that must include a consideration of the implications and the possible 'solutions' relating to facilitating mobility of EU workers, such as the enhancement of cross-border portability of supplementary pension rights.

[78] Although the OMC on pensions (Joint Report of the SPC and the Economic Policy Committee Council of the European Union (2001), 'Quality and viability of pensions: Joint report on common objectives and working methods in the area of pensions', 14098/01, SOC 469/ECOFIN 334 and in the streamlined framework, see COM (2005) 706 final) has not been used directly to promote portability of supplementary pensions and rights, this issue can be related to some of the aims the OMC sought to achieve in the area of pensions, like the promotion of the affordability and the security of funded and private schemes to achieve safe and sustainable pensions. Therefore, the OMC could have provided, in the area of supplementary pensions, and a means of exchanging best practices and to allow, through the collaboration of the Member States at national and EU level, the mobilising of all relevant actors and the gradual adjustment of national rules, for some of the obstacles impeding cross-border portability of supplementary pension rights to be addressed. See for example the National Strategy Reports of 2002 (Portugal, Spain, Italy and Belgium) and 2005 (Denmark, Netherlands, United Kingdom and Germany) where reference to measures that may be related to and affect such portability was made.

10

Defining Pollution Down: Forestry, Climate Change and the Dark Figure of Carbon Emissions[1]

MARK HALSEY

MY AIM IN this chapter is to examine the relationship between climate change, carbon sinks and forest logging and to focus particularly on the dark figure of carbon emissions from the forestry sector. In the first section, discussion extends broadly to the global context and specifically to the rate and scale of forest use (and abuse) in Australia. Despite its 'clean green' image, Australia continues to fell large areas of non-plantation native forest each year for construction purposes and for the international woodchip market. Although the Australian forest industry has access to just 6 per cent of all native forest, 80 per cent of tall open forest lies within that smaller area. Large areas of these tall open forests have recently been declared the most carbon dense in the world. The social and ecological value attached to carbon dense forest is only just starting to be weighed against their value as a timber, pulp and paper resource. How this debate is framed has profound consequences for climate change mitigation and the integrity of carbon sinks. Accordingly, in the second part of the chapter, I draw on the work of Michel Foucault and Paul Virilio in order to analyse why pollution from the forestry sector is defined down and how this impacts on national and international carbon emissions. In concluding, I briefly survey the criminological implications of the foregoing analysis.

[1] Earlier versions of this chapter were presented at 'The Legal and Criminological Consequences of Climate Change' workshop, International Institute for the Sociology of Law, Oñati, Spain 29–30 April, and at The 2010 Australian and New Zealand Critical Criminology Conference, Sydney University, 1–2 July 2010.

CARBON RESERVOIRS, GREENHOUSE GASES (GHGs) AND GLOBAL WARMING

In order to understand global warming it is important to know something about how, under 'natural' conditions, carbon is sequestered and released on earth. Globally, there are four main 'carbon reservoirs': the *atmosphere* (the air surrounding Earth), *lithosphere* (all inorganic matter such as rocks and minerals constituting the Earth's crust from the highest mountain to deepest seabed), *hydrosphere* (all water on or near the surface of Earth), and *biosphere* (all organic matter).[2] The vast bulk (99.9994%) of carbon exists in the lithosphere and this amount remains more or less constant even when measured over millions of years. It is only a very small fraction of the overall carbon reservoir that is conducive to change on time-scales that matter to human and most forms of non-human life.[3] But there is increasingly little doubt that what happens to the carbon flows in the six ten thousandths of 1 per cent of the entire carbon stock is critical to life on earth—at least in terms of the way we currently experience and understand such life.

The principal gas produced as a result of any interruption to carbon-based bodies (whether living or dead biomass or fossil fuels) is carbon dioxide. According to the Intergovernmental Panel on Climate Change (IPCC), it 'is the most important anthropogenic GHG [greenhouse gas]. [Most critically]... annual emissions [of CO_2] have grown between 1970 and 2004 by about 80%, from 21 to 38 gigatonnes (Gt), and [in 2004, CO_2] represented 77% of total anthropogenic GHG emissions...'.[4] Taking

[2] Like most categories, they are imperfect. Some organic matter (fish, seaweed), for instance, exists in the hydrosphere. And elements of the hydrosphere are found in the biosphere (water content in vegetation being a good example of such). To avoid confusion, another way of thinking about the distribution of carbon is to conceptualise it in terms of the 'CO_2 in the atmosphere, [the] organic forms of carbon in the ocean and terrestrial ecosystems (including litter and soil) and [the] geological stores of fossil vegetation in the earth's crust': RJ Keenan, 'Historical Vegetation Dynamics and the Carbon Cycle: Current Requirements and Future Challenges for Quantifying Carbon Fluxes in Australian Terrestrial Ecosystems' (2002) 50 *Australian Journal of Botany* 533, 534. One can simplify matters even further by dividing the carbon reservoir into two very broad categories: namely, 'Fossil carbon (eg coal, oil, natural gas, limestone)' and 'Living carbon (eg forests, soil carbon, marine ecosystems)': S Weaver, 'Forest Conservation and Climate Change Policy' (2006/07) *Pacific Ecologist* 21, 22. The amount of carbon estimated to be stored in each reservoir is as follows: 720 Gt in the atmosphere; 2000 Gt in the terrestrial biosphere (living and dead biomass); 4130 Gt in fossil fuels (including 3510 Gt of coal) 38 400 Gt in the ocean; 75 million Gt in the lithosphere (60 million Gt in sedimentary carbonates and 15 million Gt in kerogens): P Falkowski, RJ Scholes, E Boyle, J Canadell, D Canfield, J Elser, N Gruber, K Hibbard, P Hogberg, S Linder, FT Mackenzie, B Moore III, T Pedersen, Y Rosenthal, S Seitzinger, V Smetacek and W Steffen, 'The Global Carbon Cycle: A Test of Our Knowledge of Earth as a System' (2000) 290 *Science* 291, 293.

[3] See R Garnaut, *The Garnaut Climate Change Review: Final Report* (Melbourne, Cambridge University Press, 2008) 36.

[4] Intergovernmental Panel on Climate Change, *Climate Change 2007: Synthesis Report* (Rome, 2007) 36.

a much longer view, 'Analyses of ice cores strongly suggests that over the past 420,000 years, the climate system has operated within a relatively constrained domain of atmosphere CO_2 and temperature'.[5] This, of course, is no longer the case. Indeed, it is the speed and scale of CO_2 emissions that should be at the heart of current climate change debates. The IPCC's fourth and most recent assessment report on climate change states, 'The global atmospheric concentration of CO_2 increased from a pre-industrial value of about 280 parts per million to 379 parts per million in 2005'.[6] In the last 250 years, in other words, CO_2 emissions have increased by a minimum of 27 per cent and by as much as 47 per cent.[7]

Accumulation of greenhouse gases in the atmosphere occurs when the Earth's carbon sinks fail to absorb such gases sufficiently quickly (ie at a rate which maintains atmospheric temperatures at levels conducive to the long-term reproduction of human and nonhuman needs).[8] One major impact of GHG accumulation is that 'mean surface temperature is projected to increase by 1.4 to 5.8°C between 1990 and 2100, with this variation largely dependent on success of measures to reduce future increases in greenhouse-gas concentrations'.[9] The IPCC has remarked 'that climate sensitivity [ie the impact on climate resulting from GHG concentrations] is *likely* to be in the range of 2 to 4.5°C with a best estimate at 3°C, and is *very unlikely* to be less than 1.5°C' over the next century.[10] What this means is that global temperatures (even in light of an immediate cessation of all GHG emissions) will almost certainly continue to rise over the coming century (and

[5] Falkowski *et al*, above n 2, 291.

[6] IPCC, above n 4, 37.

[7] Falkowski *et al*, above n 2, 291.

[8] Falkowski *et al* write 'Our analysis ... shows that although natural sinks can potentially slow the rate of increase in atmospheric CO_2, there is no natural savior waiting to assimilate all the anthropogenic CO_2 in the coming century. Although on geological time scales the anthropogenic emission of CO_2 is a transient phenomenon ... it will affect Earth's biogeochemical cycles for hundreds of years' Falkowski *et al*, above n 2, 295. In 2004, 49 Gt of GHG were emitted globally as a result of anthropogenic activity. The majority (77% or 38 Gt) of such emissions took the form of CO_2. Of all CO_2 emissions, 17.3% stemmed from 'deforestation [and] decay of biomass', 56.6% from 'fossil fuel use' and 2.8% from 'other' sources. In terms of emitting sectors, 25.9% of GHG in 2004 were related to energy supply, 19.4% from industry, 17.4% from forestry, 13.5% from agriculture, 13.1% from transport, 7.9% from residential and commercial buildings, and 2.8% from the waste and wastewater sector. By volume, total global GHG emissions from the forestry sector was 8.5 Gt and of this, 6.57 Gt (or 77%) was CO_2: IPCC, above n 4, 36. Based on data from its Special Report on Emissions Scenarios published in 2000, the IPCC 'project[s] an increase of baseline global GHG emissions by a range of 9.7 to 36.7 $GtCO_2$-eq (25 to 90%) between 2000 and 2030': quoted in IPCC, above n 4, 44.

[9] Keenan, above n 2, 534.

[10] IPCC, above n 4, 38. To clarify, 'very unlikely' is defined by the IPCC as less than a 10% chance of an event occurring. 'Likely' is defined as a greater than 66% chance of the predicted outcome eventuating: IPCC, above n 4, 27.

perhaps beyond).[11] It is therefore a question of what kinds of action, if any, might help to ensure that global warming is constrained to the lowest possible margin (as near as practicable, as would seem inevitable, to 1.5°C). Where the average temperature increases to more than 1.5°C there are, to state it mildly, colossal implications for biodiversity and global ecosystemic integrity. As noted by the IPCC, 'Climate change is *likely* to lead to some irreversible impacts. There is *medium confidence* that approximately 20 to 30% of species assessed so far are *likely* to be at increased risk of extinction if increases in global average warming exceed 1.5 to 2.5°C (relative to 1980–1999). *As global average temperature increase exceeds about 3.5°C, model projections suggest significant extinctions (40 to 70% of species assessed) around the globe*'.[12] Clark and York,[13] quoting John Foster's *Ecology Against Capitalism*, remind that 'an increase of 4°C "would create an earth that was warmer than at any time in the last 40 million years"'. Alarmingly, fossil fuels are projected to maintain their dominant position in the global energy mix to 2030 and beyond. Furthermore, CO_2 emissions from all energy use are projected to grow 40 to 110 per cent between 2000 and 2030.[14] Again, the projected substantial increase in CO_2 over coming decades casts climate change mitigation as the seminal challenge of the twenty-first century. In the context of much nay saying over climate change from various quarters, Pittock[15] makes the critically import distinction between 'climate variability' (measured in years) as against 'climate change' (measured across centuries). 'Climate change', in short, 'is ... a longer-term trend superimposed on natural decade-to-decade variability'.[16]

Carbon sinks—particularly forests (planting new forests and/or ensuring that current forests are protected from becoming net carbon sources)—have been earmarked as one major way to mitigate the effects of climate change. In October 2009, the Wentworth Group of Concerned Scientists (WGCS) remarked, 'The science now tells us that it will be next to impossible for nations to achieve the scale of reductions required in sufficient time to avoid dangerous climate change unless we also remove carbon from the atmosphere and store it in vegetation and soils'.[17] To understand how carbon sinks and forestry fit into this challenge it is important to recall that global warming has resulted from a chronic (ie persistent, long-term) imbalance between the

[11] See B Pittock, 'The Science of Climate Change: Knowledge, Uncertainty and Risk' in C Lever-Tracey (ed), *Handbook of Climate Change and Society* (London, Routledge, 2010) 22–23.

[12] IPCC, above n 4, 54. Emphasis added.

[13] B Clark and R York, 'Carbon Metabolism: Global Capitalism, Climate Change, and the Biospheric Rift' (2005) 34 *Theory and Society* 391, 392.

[14] IPCC, above n 4, 44.

[15] Pittock, above n 11, 13–14.

[16] Ibid 14.

[17] Wentworth Group of Concerned Scientists, 'Optimising Carbon in the Australian Landscape' (www.wentworthgroup.org, Sydney, October 2009) 1.

production of GHGs and the rate at which the planet can sequester such gases within various sinks. As previously mentioned, the GHG of predominant concern is CO_2. Too much residual CO_2 in the atmosphere changes the long term and relatively stable effects of the process known as radiative forcing. The IPCC remark that '*Radiative forcing* is a measure of the influence a factor has in altering the balance of incoming and outgoing energy in the Earth-atmosphere system and is an index of the importance of the factor as a potential climate change mechanism'.[18] In terms of climatic effects, higher rates of radiative forcing make an already bad situation exponentially worse. In particular, 'Warming reduces terrestrial and ocean uptake of atmospheric CO_2, increasing the fraction of anthropogenic emissions remaining in the atmosphere. This *positive carbon cycle feedback* leads to larger atmospheric CO_2 increases and greater climate change for a given emissions scenario…'.[19] Trying to cool the planet in the midst of large-scale radiative forcing is akin to opening the windows of a house to take advantage of a cool breeze while running a heater on high in each room. Obviously, this would be a pointless exercise. And so, arguably, is hoping that carbon sinks can do the job they have done for thousands of years in the context of exponentially higher residual GHGs in the atmosphere (and, for that matter, the oceans).[20] With this information in mind I want now to say something about the complex but critically important relationship between forest logging (more commonly, forest harvesting) and climate change.

FORESTS, HARVESTING AND CLIMATE CHANGE

The combination of further and dramatic increases in CO_2 over coming decades coupled with the concomitant impacts of radiative forcing (which ensures that global warming would continue even if all CO_2 emissions were to cease immediately), means that forests have become terrains of central importance for reducing the amount of atmospheric CO_2. This is chiefly because, unlike other carbon sinks—such as the ocean or sedimentary rocks—they can be relatively easily and comparatively quickly managed in ways that are carbon negative. At least that is the idea. The Food and Agriculture Organisation (FAO) of the United Nations has noted that 'carbon stocks in forest biomass decreased by an estimated 0.5 Gt annually during the period 2005 [to] 2010'.[21] This is the equivalent of foregoing the

[18] IPCC, above n 4, 36 at note 4. The IPCC determines 'radiative forcing values … for changes relative to preindustrial conditions defined at 1750 …'.

[19] Ibid 38. (emphasis added)

[20] Presently, 'The radiative forcing of the climate system is dominated by the long-lived GHGs …', particularly, as mentioned, carbon dioxide: IPCC, above n 4, 36.

[21] Food and Agricultural Organisation of the United Nations, *Global Forest Assessment 2010: Key Findings* (Rome, FAO, 2010) 4.

capacity for such forest to sequester around 1.835 Gt of CO_2 (or 4% of global annual emissions) each year. Put another way, the CO_2 sequestration capacity of the 65 million hectares of forest lost to deforestation from 2005 to 2010 could have more or less off-set Australia's total greenhouse gas emissions for the next 20 years.[22] But my chief concern is not deforestation (in the technical sense of that term). Instead I want to focus on the carbon costs associated with so-called 'sustainable' forest harvesting—costs which are typically viewed as minimal or even carbon negative. And I want to look specifically at the Australian context.

In order to effectively examine the issue of forest carbon sinks it is important to give an account of global forest cover and associated carbon stores. The Forestry Department of the FAO defines forest as 'Land spanning more than 0.5 hectares with trees higher than 5 meters and a canopy cover of more than 10 percent, or trees able to reach these thresholds *in situ*. It does not include land that is predominantly under agricultural or urban land use'.[23] Under this definition, countries are able to count clearfelled areas (ie vast tracts temporarily devoid of forest) or selectively logged areas as forest as well as areas replanted with exotic tree species or single tree species but are excluded from counting areas turned over to non-forestry activity. In 2005, total forest area for the 174 countries reporting such data to the FAO was 3.756 billion hectares.[24] Of this, and for the 172 reporting countries, 1.281 billion hectares was 'designated primarily for production'.[25] On a country by country basis, Australia has the sixth largest area of forest in the world (164 million hectares)[26] and, globally, 'the ten most forest-rich countries account for two-thirds of the total forest area'.[27] Just 13.1 per cent of the 163,678,000 hectares of forest in Australia has conservation designated as its primary function. Production is the primary designated function in 8 per cent of the total area of forest, and 77.6 per cent of forest is designated as multiple purpose.[28] The area of primary (or predominantly undisturbed) forest was estimated globally to be 1.337 billion hectares or 36 per cent of the total area of forest reported by the 174 countries submitting primary

[22] Calculated on the basis of Australia's current annual emissions rate of 553 Mt CO_2-e (million tonnes of carbon dioxide equivalent): see FAO, above n 21, 3; Department of Climate Change, *Australia's National Greenhouse Accounts, National Greenhouse Gas Inventory, Accounting for the Kyoto Target, May 2009* (Canberra, Commonwealth of Australia, 2009) 1.

[23] Food and Agricultural Organisation of the United Nations, *Global Forest Resources Assessment Update: Terms and Definitions (Final Version)* (Rome, FAO, 2004) 16.

[24] These countries hold 93% of the world's forests: Food and Agricultural Organisation of the United Nations, *Global Forest Resources Assessment 2005* (Rome, FAO, 2006) 25.

[25] Ibid 79.

[26] Ibid xiii.

[27] Ibid 13.

[28] Ibid 212.

forest data.[29] Although it ranks highly in terms of forest area, Australia has very little remaining primary forest—just 5.233 million hectares (or 3% of its total forest area).[30] But as has very recently been determined, Australia does have the most carbon dense forests in the world.[31] The implications of this are profound and I will return to the issue shortly.

Australian forests hold an estimated 16.05 Gt of carbon (5.8Gt in living above-ground biomass (vegetation); 2.5Gt in living below ground biomass (roots); 2.2Gt in dead wood;[32] 2.7Gt in debris/litter;[33] and 5.5Gt in forest soil (soil organic carbon)).[34] The 16.05Gt total carbon content of Australian forests is undoubtedly a serious underestimate as there is 'substantial spatial variation in total biomass' across different stands of forested terrain[35] and quantities must ultimately be determined in relation to quite localised 'environmental conditions… life history and morphological characteristics of tree species… and the impacts of natural disturbance such as fire and land-use history'.[36] As Keith *et al*[37] continue, 'It is the interaction and feedbacks among these factors that influence vegetation community dynamics and ultimately lead to very high carbon densities'. The IPCC has classified Australian forests under the global biome of temperate forest. Work undertaken in *primary* temperate forests of Victoria and Tasmania have shown that their carbon density is the highest of any forests globally, yielding an average of '640 t C ha^{-1} of total carbon (biomass plus soil…) [and] 360 t C ha^{-1} of biomass carbon (living plus dead biomass…)'.[38] Areas of mountain ash in Victoria's Central Highlands have recently proved to be the most carbon dense of all forest stands 'with an average of more than 1200 t C ha^{-1} and maximum of over 2000 t C ha^{-1}'.[39] The authors of the study specifically draw attention to the implications of current IPCC values

[29] Ibid 41. The FAO defines primary forest as 'Forest … of native species, where there are no clearly visible indications of human activities and the ecological processes are not significantly disturbed'; FAO, above n 24, 171.

[30] FAO, above n 24, 236.

[31] H Keith, B Mackey and D Lindenmayer, 'Re-Evaluation of Forest Biomass Carbon Stocks and Lessons From the World's Most Carbon Dense Forests' (2009) 106 *Proceedings of the National Academy of Sciences of the USA* 11635.

[32] FAO, above n 24, 266.

[33] WGCS, above n 17, 4.

[34] Montreal Process Information Group for Australia, *Australia's State of Forests Report: Five-Yearly Report 2008* (Commonwealth of Australia, Canberra, 2008) 118. Globally, as of 2005, '[t]he total carbon content of forest ecosystems … [was] … 638Gt': FAO, above n 24, 35. This was composed of 283 Gt within living forest biomass, 38 Gt in deadwood, and, importantly, 317 Gt within forest soils calculated to a depth of 30 cm: FAO, above n 24, 34–35.

[35] Keith *et al*, above n 31, 11 636.

[36] Ibid 11637.

[37] Ibid.

[38] B Mackey, H Keith, SL Berry and DB Lindenmayer, *Green Carbon: The Role of Natural Forests in Carbon Storage* (Canberra, ANU E Press, 2008) 6.

[39] Ibid.

and to the paucity of assessment at the local or stand level of Australian forests. As they remark,

> the IPCC default values for temperate forests are a carbon stock of 217 t C ha^{-1} of total carbon, 96 t C ha^{-1} of biomass carbon, and an NPP [net primary productivity] of 7 t C ha^{-1} yr^{-1}. The IPCC default values for total carbon are approximately one-third, and for biomass carbon approximately one-quarter that of the average values for south-eastern Australian eucalypt forests, and one-twentieth of the most biomass carbon dense eucalypt forests. We calculate the total stock of carbon that can be stored in the 14.5 million hectares of eucalypt forest [equivalent to half of all remaining tall and medium open eucalypt forest] in our study region is 9.3 Gt(1), if it is undisturbed by intensive human land-use activities; applying the IPCC default values would give only 3.1 Gt.[40]

Critically, Mackey *et al*[41] also point to the value of local field-based (as opposed to simulated) assessment of the carbon within forests, 'The difference in carbon stocks between our estimates and the IPCC default values is the result of us using local data collected from natural forests not disturbed by logging'. The IPCC also ascribes an average soil organic carbon load of 101 tonnes per hectare to the Oceania region (which includes Australia).[42] This underestimates the soil organic content (SOC) load for much of Australian forests—especially primary forest. For instance, selected stands of tall open forest in southwest Western Australia and Tasmania have a SOC load of 123 and 271 tC·ha^{-1}.[43] Stands of forest in south eastern Australia have a SOC load up to 2000t/ha and regularly reach 350 t/ha.[44] In such light, and working with the conservative estimate of a total carbon content of 16.05 Gt, Australian forests hold at absolute minimum around 3 per cent of all carbon in forests globally as at 2005. This carbon stock functions as a sink for 60 Gt of CO_2 (or just over one year of total global GHG emissions from all sources) which would otherwise be released into the atmosphere. A more precise mapping would no doubt result in a substantially higher carbon load for Australian forests.[45] An important question therefore is: What does forestry do to or take from this potential sink on an annual basis?

— Removal of biomass

The amount of roundwood removed from Australian forests in 2005/06 was 29.7 million cubic metres (26.7 million cubic metres of industrial

[40] Ibid 6–7.

[41] Ibid 7.

[42] FAO, above n 24, 34.

[43] C Dean and G Wardell-Johnson, 'Old-Growth Forests, Carbon and Climate Change: Functions and Management for Tall Open-Forests in Two Hotspots of Temperate Australia' (2010) 144 *Plant Biosystems* 180, 182.

[44] Mackey *et al*, above n 38, 24.

[45] This is especially so given that just 9% of the total area of forest has, as previously mentioned, been recently found to store 9.3 Gt of carbon—only marginally less than the *entire* carbon stock of Australian forests depicted in the FAO Global Forest Assessment 2005 report.

roundwood and 3 million cubic metres of firewood)[46] and was valued at US$1.178 billion.[47] This is clearly a very small fraction of global production[48] but significant tracts of carbon dense native forest were (are) clearfelled or selectively logged to produce such wood. For instance, from 1 July 1997 to 31 December 2003, the small island state of Tasmania removed 312,197 logs from forest classed as 'Native Forest Mature Old-Growth'.[49] This amounted to 806,120 cubic metres of sawlogs of various grades.[50] From the same category of forest, over the same period, a further 5,027,164 cubic metres was removed as pulpwood.[51] Nearly two-thirds (65%) of the volume of logs greater than 85 cm in diameter came from Native Forest Mature Old-Growth.[52] Around 400 such logs[53] in 2001, and 150 logs in 2004, appeared to have had a diameter greater than 1.5 m.[54] More generally, almost 8.6 million cubic metres of roundwood was harvested from non-plantation native forest in Australia during 2005/06.[55] A further 3.8 million tonnes was harvested from plantation eucalypts in the same period.[56] The 12.4 million tonnes of roundwood from hardwood forests equates to 14.9Mt of CO_2 emissions in 2008—a figure reasonably consistent with the estimate given in Australia's State of the Forests Report 2008.[57] This, though, grossly underestimates actual emissions because, by all accounts, it is an estimate based solely on merchantable roundwood quantities removed (more on this shortly).

— Area of forest logged

The amount and kind of statistical information on the area and types of Australian native forest subject to harvesting in any year vary from state to

[46] FAO, above n 24, 284.

[47] Ibid 302.

[48] The total amount of roundwood removed from forests globally in 2004 was 3.4 billion cubic metres (1.623 billion cubic metres of industrial roundwood and 1.777 billion cubic metres of fuelwood). This had an estimated value of US$64 billion: FAO, above n 24, 110–11.

[49] B Felmingham, M Farley, G Lancaster and C Farley, *Impact of the Policy to Cease Clearfelling of Old Growth Forests in 2010: An Overview of Productivity, Financial and Employment Impacts, A Report Prepared for the Tasmanian Forest and Timber Industries* (Hobart, Symetrics Business Intelligence, 2004) 53.

[50] Ibid 16.

[51] Ibid 19.

[52] Ibid 23.

[53] Logs are necessarily smaller in diameter than the tree from which they are cut.

[54] Felmingham *et al*, above n 49, 57.

[55] Australian Bureau of Agricultural and Resource Economics, *Australian Forest and Wood Products Statistics: September and December Quarters 2008* (Canberra, Commonwealth of Australia, 2009) 12.

[56] Ibid 'The vast majority of carbon (99.2%) in Australian forests is in natural, mainly hardwood, forests (both commercial and conservation forests), with the balance in plantations': Forest and Wood Products Research and Development Corporation, Forests, Wood and Australia's Carbon Balance (Australian Government, Canberra, undated) 2.

[57] MPIGA, above n 34, 114–15.

state. Such information, however, is critical to discerning the relative annual damage done to the forest carbon sink (which includes soils, above ground and below ground biomass, deadwood and litter within forests) and to the sequestration capacity forgone over many decades as a result of such logging. The following statistics give some idea of the area logged in recent times across Australia. In Tasmania during 2008/09, 27,291 hectares of non-plantation native forests was logged –14,601 hectares was clearfelled (66% or 9,651 ha from private land) and 1,922 hectares was turned over to another use (ie not reseeded or planted).[58] In New South Wales, 31,252 hectares of native forests was logged during 2008/09 (Department of Primary Industries 2009: 7).[59] In Victoria, 9,470 hectares of native forest was logged during 2005/06.[60] In Queensland—a state where clearfelling does not occur—up to 50 per cent of the basal area of suitable trees within a given 120,000 hectare area of native forest is selectively logged each year (with 90% of such logging occurring in Spotted Gum and Ironbark forests).[61] In Western Australia, 'around 0.46%', or 11,316 hectares of the total area of native forest (2.46 million hectares) 'is [formally] available for harvest each year'.[62] The total net area of coupes scheduled for logging in 2010 in that state is 36,290 hectares (which includes a proportion of coupes carried over from previous years). This is projected to yield an estimated 262,650 cubic metres of sawlogs and does not account for the pulplog volume which will also be removed.[63]

In all, then, and, very conservatively, around 120,000 hectares of native forests in Australia is directly impacted by intensive logging on an annual basis (a figure which excludes the impact on surrounding forest area in terms of road building or thinning activities in other parts of the forest estate). It is

[58] Forestry Practices Authority, *Annual Report 2008/09* (Tasmanian Government, Hobart, 2009) 11.

[59] This is not the same as saying that all trees within this area were removed. In fact NSW has quite a varied approach to harvesting dependent on forest type. But the most common method—'single tree selection'—permits up to 40% of the pre-harvested basal area of each coupe to be removed. In 2008/09 this method was applied to 13,894 hectares of non-plantation native forest. Clearfelling (also called the 'alternate coupe' approach) was used for 1,162 hectares, 'group selection' across 2,207 hectares, and 'commercial thinning' occurred in 8,861 hectares of non-plantation native forest: Department of Primary Industries, Forests NSW Annual Report 2008–09: Social, Environmental and Economic Performance (Government of New South Wales, West Pennant Hills, 2009) 7.

[60] Department of Sustainability and Environment, *Victoria's State of the Forests Report 2008* (Victorian Government, Melbourne, 2009) 16.

[61] Queensland Department of Environment and Resource Management, www.derm.qld.gov.au/forests/saleable_products/forest_industry.html, accessed 29/06/10; Queensland Department of Environment and Resource Management, www.derm.qld.gov.au/forests/saleable_products/timberproduction/harvesting.html, accessed 29/06/10; and in conversation (Department of Environment and Resource Management, 26 March 2010).

[62] Western Australia Forest Products Commission, www.fpc.wa.gov.au/content_migration/native_forests/protection.aspx, accessed 13/04/10.

[63] Western Australia Forest Products Commission, www.fpc.wa.gov.au/content_migration/native_forests/harvest_plans/2010/Default.aspx, accessed 13/04/10.

an area which continues, on the best available evidence, to rely in substantial measure on old-growth and mature or over-mature forest—forest which, not incidentally, has the highest carbon load across all growth stages.[64] In attempting to deduce the annual emissions from harvesting of Australian native forests[65] it has been reliably determined that the 2500 hectares of old-growth forest harvested in Tasmania during 2004/05 had a carbon density of 358 t/ha and emitted 895 000 tonnes of CO_2-e.[66] Conservatively, old-growth forests account for 25 per cent of all native forest harvested annually in Australia[67] with the remainder being drawn from mature and over-mature eucalypt forests more than 110 years old (exhibiting a very high carbon load) and from regrowth forests recovering from previous serious disturbance. Using a carbon density of 350 t/ha for 25 per cent of the annual harvested area (old-growth), and a carbon density of 200t/ha for the remaining area, this yields an annual CO_2 emission level of 105.4 Mt.[68] This is equivalent to almost one fifth of Australia's annual GHG emissions and does not take into account emissions from machinery and infrastructure associated with forestry activities and excludes any calculation (most likely to be of a very considerable nature) of the amount of carbon released through disturbance and compaction of forest soils. It also precludes the loss to the carbon store as a result of native forest being converted to plantation which is thought to be in the order of 40 per cent.[69] The figure of 105 Mt

[64] See Dean and Wardell-Johnson, above n 43.

[65] This is an incredibly difficult task made so by the paucity of metrics and databases which record precisely this information. There are no known comprehensive or, for that matter, indicative aggregated figures which show the amount of CO_2-e released from the biomass and soils of 'natural' and 'industrialised' forests as a result of logging activity. Mackey *et al* further note that 'We cannot find any ... comprehensive accounts of the grey carbon ['the carbon stored in fossil fuel'] emitted from commercial logging and wood-products manufacturing inclusive of all stages in the product life cycle: forest management, harvesting, transportation and manufacturing': Mackey *et al*, above n 38, 11, 18.

[66] Dean and Wardell-Johnson, above n 43, 187–88.

[67] Much of this occurs in Tasmania due to the comparatively intact nature of forests in that state. Victoria, by contrast, reports that just 50 ha of old-growth forest was logged in 2005/06—down from 730 ha in 2001/02: Department of Sustainability and Environment, above n 59, 19. But see discussion of 'negligibly disturbed forest' further in this chapter.

[68] Based on 10.5 Mt C in old-growth (30,000 ha / 25% of harvested area) producing 38.8 Mt of CO_2 and 18 Mt C for mature/over mature/ regrowth (90,000 ha / 75% of harvested area) producing 66.6 Mt of CO_2.

[69] 'Commercial logging changes the age structure of forests so that the average age of trees is much younger. The result is a significant (more than 40 percent) reduction in the long-term average standing stock of biomass carbon compared with an unlogged forest ... Plantations are designed to have all of their above-ground biomass removed on a regular basis. The rotation period between harvests varies from 10 to 70 years globally, depending on species and commercial purposes ... The carbon stock of forests subject to commercial logging—and of monoculture plantations in particular—will therefore always be significantly less on average than the carbon stock of natural, undisturbed forests': Mackey *et al,* above n 37, 17. Dean and Wardell-Johnson found that 'Conversion of "pristine" [mountain ash] forest [in Tasmania] to plantation reduced SOC by 25% to 0.45m depth after several plantation cycles': Dean and Wardell-Johnson, above n 43, 183.

from harvesting of native forest sits in stark contrast with the 14.1 Mt of CO_2 harvest-related emissions officially reported in Australia's most recent State of the Forests Report which is based only on the carbon loss attached to roundwood removals.[70] It is worth quoting in full the relevant passage from this report:

> Harvesting from managed native forests (that is, forests subject to harvest and regrowth from prior harvest) averaged around 11.8 million cubic metres of roundwood per year in the period from 2001 to 2005. About 3.8 million tonnes of carbon—equivalent to about 0.06% of the total carbon stored in native forests and 14.1 million tonnes of carbon dioxide—was removed. This small removal was exceeded by new growth: about 43.5 million tonnes of carbon dioxide, equivalent to about 11.9 million tonnes of carbon, was estimated to have been sequestered in managed native forests in 2005 after taking into account the decay of slash produced during harvesting ... Therefore, yearly sequestration of carbon in 2005 was about three times yearly emissions and removals arising from managed native forests subject to harvest and regrowth from prior harvest. Because the carbon stored in native forest biomass, including in nature conservation reserves, is 20 times more than that contained in forest plantations, wood in service and wood in landfill combined, it has a dominant effect on the overall amount of carbon stored by the forest sector. Harvesting (<1% of the native forest outside nature conservation reserves annually), planned and unplanned fires ... and other periodic and infrequent disturbances therefore had little impact on national average total native forest carbon stock over the longer term.[71]

The key point is that the 43 Mt of CO_2 annually 'sequestered in managed native forests' makes sense if one calculates carbon loss (3.8 million tonnes) *only* in relation to roundwood removals—that is, by the merchantable timber made available to various markets.[72] *By my very conservative estimate, harvesting activity in native forest is, therefore, outstripping net sequestration by remaining forest by a ratio of at least 2.5 to 1* (105 Mt emitted as against 43 Mt sequestered). There is a widely held view that forestry is a

[70] MPIGA, above n 34, 114–15.

[71] MPIGA, above n 34, 114–15.

[72] This is precisely how statistics given in *Australia's Fifth National Communication on Climate Change: A Report Under the United Nations Framework Convention on Climate Change 2010* can determine 'forested land' (with the exception of 1991, 2003, 2004 and 2006) as a *net* carbon sink for the period 1990 to 2007 (with the highest sink rate given as 95.7Mt CO_2 in 2005): Department of Climate Change, *Australia's Fifth National Communication on Climate Change: A Report Under the United Nations Framework Convention on Climate Change 2010* (Canberra, Commonwealth of Australia, 2010) 181. Indeed, in terms of the land use, land use change and forestry sector, it is 'grassland' which proves to be the only consistently positive emitter of GHGs with emissions of 282 Mt in 2007 alone (predominantly from the effects of land clearing), 181. In 2007, net GHG emissions in Australia excluding the land use, land use change and forestry sector was 541 Mt CO_2-e. *Including* the sector puts total emissions for 2007 at 826 Mt CO_2-e (an increase of 34%), 32. This net emission level is increased further still when one begins to make even modest allowances for the amount of additional aboveground biomass disturbed during harvesting operations.

carbon negative activity due to the 'vigorous' growth of plantation forests and because wood products continue to store large amounts of carbon for many years. This is incorrect. Industrialised forests have nothing close to the sequestration rate and long-term carbon store of the mature, over-mature and old-growth forests they replace. Further, only around 3 to 5 per cent of the total forest above ground biomass is recovered as sawn product and of this small fraction, only 1 per cent could be said to store its original carbon load (in the form of high quality hardwood products such as floors, tables and chairs) over the longer term (longer than, for instance, the 50 to 80 year rotation cycles of industrially managed forests).[73] In other words, 95 per cent of the biomass within harvested areas returns to the atmosphere in the form of CO_2 either through burning of forest 'debris' within coupes or through conversion to woodchips for pulp and paper products.[74] In addition to all this, there are profound implications for the long-term eco-systemic health of forests—indeed what counts as a forest—in the context of converting large tracts of multi-age and intensely biodiverse forests into managed industrial zones. It is well established that eco-simplification and eco-fragmentation of forests are two sure-fire ways to reduce the resilience of particular biomes over the longer term—a resilience that is urgently needed in large quantum in the context of climate change.

ANALYSIS

In this final section I want to explore why the logging of Australian forests is cast as sustainable and carbon negative when there is mounting evidence to the contrary. And I want to examine what this 'defining down' of pollution means in an international context for climate change mitigation. As has already been shown, there are an array of discourses ('truth games') which construct climate change and its solutions in various ways and which require us to think through such matters as the relationship between (liberal democratic forms of) power and (scientifico-industrial kinds of) knowledge as well as the (biopolitical) management of risky populations (human and nonhuman). The work of Michel Foucault can certainly assist to unpack such matters. It is also important to recognise that particular technologies have irrevocably transformed not just the natural world, but the place of human beings within in it. The rate and scale of this transformation has increased exponentially over the last century or two and this has ongoing implications for climate change and its mitigation. Paul Virilio—the

[73] Dean and Wardell-Johnson put the ratio of sawlog to pulplog retrieval at four sawlogs per 100 pulplogs: Dean and Wardell-Johnson, above n 43, 188.

[74] See www.eastgippsland.net.au/?q=campaigns/carbon, accessed 14/04/10, and in conversation with Jill Redwood, 14/04/10.

seminal architect of the cultural, social and ecological effects of speed—can, I believe, assist to disaggregate some of the pressing issues here.

— Climate change as biopolitical event

Foucault has written, '[I]n every society the production of discourse is at once controlled, selected, organised and redistributed according to a certain number of procedures, whose role is to avert its powers and its dangers, to cope with chance events, to evade its ponderous, awesome materiality'.[75] This is certainly true of climate change. There may be much chatter about the issue but what counts is whether one is capable of entering into a *discourse*—a rational, technical and predominantly sutured exchange about climate change and its 'solutions'. Discourse is based as much on the power to exclude various perspectives as it on the power to include knowledge-claims.[76] Currently, the 'official' dimensions of climate change—its antecedents, its present effects, its predicated impacts—find their footing in the meta-discourse transmitted through the IPCC and have done so since the 1980s when it first placed climate change on the world stage.[77] Angela Oels has argued that 'Under the auspices of the IPCC, climate change was framed as an issue of planetary management that required natural science expertise and a technological fix on that basis'.[78] Clearly, the IPCC is but one way of thinking through and distributing statements (as opposed to chatter or noise) about climate change. But it is an undeniably important body and must therefore be reckoned with.

Foucault offers a novel way of conceiving of institutions such as the IPCC (and relevant bodies such as the Australian Government's Department of Climate Change). Firstly, he is very clear about the modern era and the features which need to inform social analysis of that era. Specifically, one needs to recognise the historical convergence of (a) a new economic order (capitalism), (b) new types of knowledge (the 'hard' and 'soft' disciplines), and (c) new forms of social control (biopower). All discourse on climate change and mitigation strategies arises within and is constrained by this triptych. The IPCC resides in the shadow of the monumental shift from feudalism to capitalism and from sovereign power (whose concern was

[75] M Foucault, 'Orders of Discourse' (1971) 10(2) *Social Science Information* 7, 8.

[76] As Foucault puts it, 'We must conceive discourse as a violence that we do to things, or, at all events, as a practice we impose upon them; it is in this practice that the events of discourse find the principle of their regularity': ibid 22.

[77] The presence of large quantities of CO_2 in the atmosphere have been known for at least a century. 'While not recognizing the potential dangers associated with increasing global temperatures, Arrhenius in 1896 noted that industrial operations were contributing to an increase in CO_2 in the natural world. We know that the quantity of CO_2 in the atmosphere "has increased 31 percent since preindustrial times" and that "half of the increase has been since 1965"': Clark and York, above n 13, 403.

[78] A Oels, 'Rendering Climate Change Governable: From Biopower to Advanced Liberal Democracies' (2005) 7 *Journal of Environmental Policy and Planning* 185, 198.

obedience of subjects within small defined territories), to disciplinary power (whose concern was the productivity of bodies within states), to biopower (whose concern is the regulation of vast populations across borders). In contrast to sovereign power and its cavalier attitude to death (or the welfare of subjects), the key aim of biopower—especially in the context of liberal democratic society—is to *enhance the conditions for life*. This is why Foucault considers 'forecasts, statistical estimates, and overall measures'[79] as well as 'the emergence of demography, the evaluation of the relationship between resources and inhabitants, the constructing of tables analysing wealth and its circulation'[80] to be so important in the modern era. Here, for the first time, governments seek to know and manage the citizenry as a 'massifying' rather than 'individualising' force.[81] Humankind has not, on this count, been human*kind* for very long.[82] A profound implication of this is that institutions tasked with managing the impact of populations (dealing with citizens as a massifying force) are still in their infancy. And yet these same institutions (national governments, international bodies) must now gain 'control over relations between the human race ... *and their environment...*'.[83] Foucault termed this challenge the 'last domain' of biopolitics—centring as it does on the entry of ecological issues into the government of (world) populations. The most important weapon for wielding effective biopolitical government—for maintaining the health and vitality of the citizenry—is knowledge of the risks posed *to* and *by* various populations. Over the last two centuries, modern governments have become incredibly adept at using various devices (demographic, agronomic, economic, medical, and so forth) to promote and preserve (human) life. Indeed human populations now prosper in circumstances where they once would have languished.[84] But this success has been overwhelmingly bad for nonhuman populations—the unremarked dimension of the biopolitical ledger. Of course, the life-cycles of nonhuman species have been intensely and extensively studied for many decades. But the knowledge generated—particularly knowledge of their rarity, their ecological and cultural signficance, and the factors impacting their decline—has tended to conflict with the ongoing biopolitical management (survival) of particular

[79] M Foucault, *Society Must Be Defended: Lectures at the College de France, 1975–76* (New York, Picador, 2003) 246.

[80] M Foucault, *History of Sexuality* 1 (London, Penguin, 1978) 140.

[81] Foucault, above n 79, 243.

[82] As Foucault writes, from the end of the eighteenth century, 'Western man was gradually learning what it meant to be a living species in a living world, to have a body, conditions of existence, probabilities of life, an individual and collective welfare, forces that could be modified, and a space in which they could be distributed in an optimal manner': Foucault, above n 80, 142.

[83] Foucault, above n 79, 244–45. Emphasis added.

[84] This is not, of course, to deny the ongoing presence and effects of famine and disease in various parts of the world.

human populations (loggers, miners, fisherpersons, heavy-industry leaders, financiers, stock-brokers, and the like). Luke captures the essence of the issue by remarking, 'As human beings began to consciously wager their life as a species on the outcomes of ... biopolitical strategies ... [strategies which aimed to guarantee longer, healthier, wealthier lives], it became clear that they also were wagering the lives of other (or all) species as well'.[85]

An inevitable component of biopolitics—of managing what Foucault[86] termed 'aleatory events' on a global scale—is that particular populations will be required to cease doing what they do in order to restore the capacity for care to be extended to '*all* living [things]'.[87] As Dillon puts it, there comes a point where 'life inimical to life [has to] be differentiated, corrected, punished and, if necessary, eliminated'.[88] This is a pivotal statement. For is this not, in effect, what the IPCC and various governments are now charged with achieving—namely, the 'differentiating' and 'correcting' of rogue forces (particular patterns of human production and consumption) warming the planet? And yet, aren't these institutions proceeding in ways which try to preserve the fundamental tenets of the system responsible for such warming—global capitalism (what Guattari calls, World Integrated Capitalism)? Is it not capitalism that sits at the heart of the process of life becoming inimical to life—of humans as a species extinguishing and decimating other species on account of being fervently wedded to individualism, technological determinism, and exponential economic growth at all costs (even at the cost of passing off piece-meal climate change mitigation measures as wholesale long-term solutions)?

The power of capital accumulation—the force of economic rationalist logic—tightly constrains what counts as an appropriate response to climate change (indeed whether climate change is perceived *as* a problem). To return to an earlier theme, this is where discourse emerges as the means to stage-manage the impact of one species on all others. One presently speaks not of capitalism but of the free-market, not of pollution but carbon credits, not of fossil fuels but of green coal, not of forest or biodiversity but of carbon sinks. The discourse of *ecological modernization*, as Luke[89] calls it, is incredibly powerful—indeed, it is the political technology by which the West hopes to 'have its cake and eat it too'. As Baldwin writes, '[T]he apprehension of knowledge about how it is that ecosystems are central to human survival (eco-knowledge) becomes a political technology through

[85] T Luke, 'On Environmentality: Geo-Power and Eco-Knowledge in the Discourses of Contemporary Environmentalism' (1995) 31 *Cultural Critique* 57, 66.

[86] Foucault, above n 79, 246.

[87] M Dillon, 'Cared to Death: The Biopoliticised Time of Your Life' (2005) 2 *Foucault Studies* 37, 44. Emphasis added.

[88] M Dillon, 'Governing Through Contingency: The Security of Biopolitical Governance' (2007) 26 *Political Geography* 41, 43.

[89] Luke, above n 85.

which geo-power is exercised'.[90] It certainly matters that 'For several hundred years, capitalism has been the global hegemonic economic system, influencing human interactions with nature'.[91] But what is also crucial is that this same 'hegemonic economic system' has generated wealth as well as poverty, social mobility as well as social exclusion, and technologies put to peaceful and well as violent use. And it is a system which has always been buoyed by the abject human and ecological costs often associated with various kinds of communism, socialism, anarchism in play at various historical junctures. Capitalism has tremendous staying power and its fundamental tenets are not going to wither any time soon—not even in the face of reflexive knowledge about its capacity to threaten the world's ecosystems. All this is a way of saying that climate change mitigation takes place within very narrow parameters and that certain things—even things believed very likely to make a substantial difference to CO_2 emissions (cessation of burning coal, cessation of logging in mature multi-age forests)—are simply off the table or positioned beyond the limits of (rational) thought. It has become forebodingly easy, in the climate of climate change, to be cast as heretical or polemical. As Foucault writes,[92]

> none may enter into discourse on a specific subject unless he has satisfied certain conditions or if he is not ... qualified to do so. More exactly, not all areas of discourse are equally open and penetrable; some are forbidden territory (differentiated and differentiating) while others are virtually open to the winds and stand, without any prior restrictions, open to all.

In terms of halting emissions from the forestry sector, there seems to be limited discursive space open to those who would propose an immediate moratorium on the logging of multi-age native forests (especially old-growth) for wood and paper products. The forest and wood products industry in Australia is a comparatively small employer, costly in terms of the trade deficit, and a considerable contributor to greenhouse gas emissions.[93] It has nonetheless withstood all kinds of pressure from environmental groups and has managed to ensure that large tracts of the most carbon

[90] A Baldwin, 'The Nature of Boreal Forest: Governmentality and Forest-Nature' (2003) 6 *Space and Culture* 415, 419.

[91] Clark and York, above n 13, 392.

[92] Foucault, above n 75, 17.

[93] According to ABARE, the total number of people employed in Australian 'forestry, logging and wood manufacturing' for 2007/08 was 76,800: ABARE, above n 55, 25. To 30 June 2009 'the annual value of turnover of Australia's forestry and forest products industries [was] $23 billion': Department of Agriculture, Fisheries and Forestry, *Australia's Forests at a Glance 2010: Data to 2009* (Canberra, Commonwealth of Australia, 2010) 1. But in GDP terms, the wood and wood products sector has made an annual loss across many decades (with, for example, $4.5 billion of such products imported against $2.3 billion exported during 2008/09). Almost $1 billion of these exports took the form of woodchips from native forests with around 50% (5.1 million tonnes) sourced from non-plantation native forest (Department of Agriculture, Fisheries and Forestry, above n 92, 3; ABARE, above n 55, 12.

dense forest are zoned multiple use (ie areas where commercial harvesting is permitted). The native forest logging industry in Australia relies on continuing access to two major forest types: tall open native forest (potential mature height of > 30 metres) and medium open native forest (potential mature height of 10 to 30 metres). There is an estimated 5.8 million hectares of tall open forest remaining in Australia (just 4% of all native forest) and a staggering 78 per cent of such falls beyond 'strict nature reserves' (75,000 ha), 'wilderness areas' (383,000 ha) or 'national parks' (879,000 ha). Nearly half the tall forests in Tasmania, for instance, are zoned multiple use.[94] There is an estimated 28.1 million hectares of medium open native forest (19% of all native forest) and just over 80 per cent of this falls outside the aforementioned categories which would prevent access by the forest industry.[95] The forest industry endlessly promotes the fact that the area of 'Public native forests where timber production is permitted (gross area)' is 9.4 million hectares[96] or just 6 per cent of all native forest in Australia. But what is rarely made clear is that the forest to which they have access includes around 4.5 million hectares (or nearly 80%) of all remaining tall open forest and the best stands within the remaining 23.2 million hectares of medium open forest.[97] Arguably, the prospect of securing sufficient terrestrial carbon sinks is undermined by an industry which is legally permitted to turn large areas of the most biodiverse and carbon dense forest on the continent into monocultures.

— Climate change and the diminution of (ecosystemic) difference

'A famous national [ie Australian] proverb dictates: "If it moves, shoot it; if it doesn't, chop it down"'.[98] There is, in other words, a historical envisioning of forests that is stubbornly tied to their removal—whether by axe, chainsaw, fire or bulldozer. To cite Griffiths again,[99] 'The historical and moral momentum for tree clearing is immense; it is an "improvement", as legislators have decreed and crown bailiffs have adjudged for a century and a half'. With the rise of large-scale conservation movements and technical advances in forest mapping such as satellite imaging and remote sensing, it is commonplace to believe that old attitudes toward 'forests as obstacles to settlement and progress' are long gone and that we now can see, categorically, what a forest is, how forests function, what quantum of wood they contain, what their growth rates are, and so forth. In fact the mode of envisioning that invests forest management in twenty-first century Australia

[94] MPIGA, above n 34, 30.

[95] See MPIGA, above n 34, 8, 22 and 24.

[96] Department of Agriculture, Fisheries and Forestry, above n 93, 2.

[97] MPIGA, above n 34, 24.

[98] T Griffiths, 'How Many Trees Make a Forest?' (2002) 50 *Australian Journal of Botany* 375, 378.

[99] Ibid 379.

is short sighted at best and blind at worst. Remarkably, only 13 per cent (14.9 million hectares) of the various growth stages of all eucalypt forest has been mapped.[100] And less than 16 per cent of the total area of native forest in Australia has been 'assessed for old-growth values'.[101] Technology, in short, has taken us *out* of the forests and *into* the lab. It is no coincidence that the authors who recently discovered the highest forest carbon load in the world did so through painstaking and lengthy fieldwork in the forests proper. To reduce a forest to its potential yield of wood products is to invoke, as Foucault put it, 'discourse as a violence that we do to things'.[102]

The issue of biodiversity depletion—the flattening of ecosystemic difference[103]—cannot be divorced from climate change. One of the great catch-cries of the forestry sector is that it is a net carbon sequester—meaning that forests disappeared through harvesting are compensated for by the 'vigorous uptake' of CO_2 in the remainder of the forest estate. This, however, is only possible on the basis that one excludes all kinds of critical variables: differences in carbon density across forest stands; impact of logging operations on soil organic carbon load; CO_2 emissions from burning 'defective' logs and other vegetation left on the forest floor; impact of road building and log haulage; CO_2 lost through conversion of timber into wood products with a very short carbon sequestration half-life; and so forth. The forestry sector has long argued that forest with 'high conservation values' have been adequately protected in national parks and the like. In fact, just 16 per cent (23 million hectares) of the entire forest estate in Australia is in conservation reserves.[104] More importantly, of the 33.9 million hectares of medium and tall open forest only 18 per cent is afforded ongoing protection from harvesting. Still, the forestry sector—indeed state and federal governments—believes that forests can be felled in good conscience because the boundary between high conservation

[100] MPIGA, above n 34, 18. Unsurprisingly, the forest type with the highest proportion of completed mapping is tall open forest (88% of total estimated area)—forest mapped no doubt in order to better forecast industrial removals and outputs than for conservation purposes: MPIGA, above n 34, 8.

[101] Ibid 2. And yet the comprehensive, adequate and representative (CAR) reserve system requires Australia to place a minimum of 60% of *existing* old-growth forest types in conservation reserves: MPIGA, above n 34, 8. Clearly, 60% of an unknown quantum is a situation open to all kinds of abuse.

[102] Foucault, above n 75, 22.

[103] Gilles Deleuze argues that life is characterised by a 'deep disparity'—an ineluctable and enduring difference: G Deleuze, *The Logic of Sense* (New York, Columbia University Press, 1990) 261. Williams, commenting on Deleuze's work, insightfully reminds that 'Difference is not a limit that stands as the origin or the end of a process of identification ... Rather, difference is that which turns all representations into illusions—identity is only a cloak thrown over deeper pure differences'. On this count, the removal and replacement of one forest by another always entails the loss of unique relationships between vegetation, soil and plant and animal communities: J Williams, *Gilles Deleuze's Difference and Repetition* (Edinburgh, Edinburgh University Press, 2003) 56.

[104] Department of Agriculture, Fisheries and Forestry, above n 93, 2.

forest and multiple use forest is a scientifically valid one. What recent work has shown is that this system of forest reserves in Australia leaves a good portion of carbon dense forests exposed to harvesting operations—that the best carbon sinks also reside within the forests given over to multiple use rather than the forests ascribed permanent conservation status. In addition to biodiversity depletion, there is, therefore, a new type of danger associated with forest activities—carbon de-densification.

— Climate change as general accident

Paul Virilio has written extensively on the relationship between technology and the unforeseen events attending particular inventions. As he writes, 'Every technology produces, provokes, programs a specific accident ... The invention of the steam engine and the locomotive was the invention of derailments. The invention of the highway was the invention of three hundred cars colliding in five minutes ... [W]e must think about both the substance and the accident—substance being both the object and its accident...'.[105] Elsewhere, Virilio remarks, 'I always try to emphasize both the invention and the accident. But the occurrence of the accident is being denied'.[106] Over the twentieth century, the invention of the chainsaw, the articulated trailer, and the dozer (footed track) combined, in the first instance, to produce the specific accident of habitat fragmentation and simplification (even if this was and continues to be called sustainable harvesting).[107] More recently, and even accepting major advances in forest mapping and ecosystemic knowledge, the application of these technologies in the tall and medium open forests of Australia are producing the specific accident of carbon deficiency or scarcity. Slowly but surely some of the most efficient and largest carbon sinks are being defiled, contributing to what Virilio has termed the 'general accident' of global warming. As he comments, 'Today, the new technologies convey a certain type of accident, one that is no longer local and precisely situated, like the sinking of the *Titanic* or the derailment of the train, but *general*, an accident that immediately affects the entire world'.[108]

[105] P Virilio, *Pure War* (New York, Semiotext(e), 1997) 38.

[106] P Virilio, *Politics of the Very Worst* (New York, Semiotext(e), 1999) 40.

[107] As an example of the effect of such technologies, 'the State forests of Croajingolong [far eastern Victoria] were, as of 1907, producing under 3% of the sawn hardwood timber in Victoria. This figure remained constant as of 1946 where such forests were used to produce 'less than five million [superfeet] per year' or 3% of the State's total hardwood output of around 160 million superfeet [citation omitted]. However, by 1969, these same forests—which had since 1770 remained predominantly free from intensive and sustained commercial exploitation—would be producing 25% or 107 million super feet per annum (about 250,000 cubic metres) of Victoria's total output of hardwood timber of around 428 million super feet (1 million cubic metres) [citation omitted]': M Halsey, *Deleuze and Environmental Damage* (London, Ashgate, (2006) 140–41.

[108] Virilio, above n 106, 12.

Speed (time) is critical in this context. The consequences of climate change are simultaneously present yet in effect displaced to future decades (and therefore in a real sense not yet real). Climate change is mass damage in slow motion. The politics of forest 'loss'—what counts as loss, or damage, or recklessness in terms of harm to carbon loads—is typified by, on the one hand, knowing that the metabolic relationship[109] between humans and Earth must be resolved, as against the perception that global warming is yet to concretely and repeatedly infiltrate our daily routines in durable and direct ways. There is much, therefore, about the clash of geological and (post)industrial speeds and rhythms that remains to be critically examined. For example, forests that have taken thousands of years to evolve give the illusion of ecological intactness even where this has been seriously assaulted. One can regrow a stand of mountain ash, English Oak, or Californian Redwoods. But no society has determined how to regrow an old-growth ecosystem. In the context of his work on cinema, Virilio comments, 'Things exist even more so because they disappear. Film is an aesthetics of disappearance staged in sequences'.[110] The forest industry has become very adept at ensuring that forests are not seen or sensed to disappear even in the context of vast tracts being removed each year for industrial purposes. We are in the bizarre and dangerous age where no forest ever disappears because it forms part of a continuous cycle of harvesting and renewal. The definition of deforestation used by the IPCC is the archetypal example of permitting disappeared forests to be permanently present and counted as such. What falls outside this vision of forests, though, is the exponential loss incurred as a result of making previously diverse and carbon dense forest conform to the demands of 80 year harvesting rotations. These rotation cycles are, arguably, the equivalent of instituting an 'aesthetics of disappearance staged in sequences' of such terrains. It is the carbon density and biodiversity of the native forest estate that is disappearing. The question is: Will (the costs of) this disappearance be noticed and arrested in a timely enough manner to avert a pronounced change in climate?

[109] 'The historic accumulation of CO_2 in the atmosphere involves the rupture of the carbon cycle. Human societies, in their metabolic exchange of materials with the environment, serve as the driving force creating a rift in the carbon cycle ... [I]t was not until the rise of capitalism, and especially the development of industrial capital, that anthropogenic CO_2 emissions greatly expanded in scale, through the burning of coal and petroleum, exploiting the historic stock of energy that was stored deep in the earth and releasing it back into the atmosphere. As a result, the concentration of CO_2 in the atmosphere has increased dramatically, overwhelming the ability of natural sinks—which have also been disrupted by anthropogenic forces—to absorb the additional carbon and leading to climate change': Clark and York, above n 13, 403.

[110] Virilio, above n 105, 22.

CONCLUDING REMARKS

The phrase 'Defining deviancy down' was devised by US Republican Senator Daniel Moynihan in the early 1990s. His intention was to give expression to the way in which particular harms were framed out of the remit of the criminal justice system. Karmen, in his appropriately pointed attack on the Senator's political/social agenda, commented that 'defining deviance down' is a new phrase designed to draw attention to old battle cries such as '"permissiveness", ... "being soft on crime" ... "moral decay"...' and the generalised need to 'restor[e] law [and] order'.[111] In this chapter, I have deliberately tried to avoid such moral posturing. Specifically, I have not advocated a 'get tough on crime' stance. Nor do I see the sense in expanding the list of things defined as criminal as a means to reign in criminal conduct. Instead, the phrase 'defining deviancy down' has functioned in this chapter as a heuristic device enabling definitions of harm (in this case, environmental harm) to be understood in relation to particular discourses—discourses which often push to the periphery the kinds of information which should be at the centre of debate. This is of course a reference to the destabilising and yet critically important role played by the dark or forgotten figure of crime—a figure which has been of concern to scholars and policy makers at least since the mid-nineteenth century.[112] The dark figure of (forestry related) carbon emissions is sustained through a complex series of rhetorical manoeuvres which purport to mitigate climate change.

Definitions of acceptable and unacceptable environmental conduct are absolutely critical in this context. The FAO[113] 'specifically excludes ... harvesting or logging' from the definition of deforestation which instead equates to 'long-term or permanent loss of forest cover' and/or conversion of forest to another use (eg pasture or agriculture). In the Australian context this conveniently excludes the annual conversion of around 120 000 hectares of non-plantation native forest into an industrial resource for the supply of timber and pulp.[114] The definition of forest degradation is also cleverly couched in terms of 'changes ... which negatively affect the structure or function of the stand or site and thereby lower the capacity to supply products and/or services'.[115] This enables forests to be extricated from their biological and geological rhythms and forced into the metrics associated with industrial time. Sustainability becomes market driven

[111] A Karmen, '"Defining Deviancy Down": How Senator Moynihan's Misleading Phrase About Criminal Justice is Rapidly Incorporated Into Popular Culture' (1994) 2(5) *Journal of Criminal Justice and Popular Culture* 99, 103–04.

[112] See A Biderman and AJ Reiss, 'On Exploring the "Dark Figure" of Crime' (1967) 374 *Annals of the American Academy of Political and Social Science* 1.

[113] FAO, above n 21, 25–26.

[114] Two-thirds (6 million cubic metres) of the total roundwood quantity extracted each year is converted to woodchips.

[115] FAO, above n 21, 26.

rather than eco-systemically based. A 'healthy' forest now equates to a uniform even-aged group of trees harvested in accordance with set rotation cycles. In this brave new world, the forest as multi-age diverse organism is anathema to the prescribed needs of industry. Primary forest—what is also termed, old-growth forest—is therefore rendered particularly problematic. Such forest contains a broad mix of trees in terms of species and age—many of which are unsuitable for milling as sawlogs (due to being 'defective', 'crooked', 'hollow', and so forth). But primary forest—especially the tall open eucalypt forests of south eastern Australia—contain some of the most prized examples of trees suited to wood products. Since at least the 1960s, this tension has resulted in many battles between loggers and conservationists.[116] Ongoing access by the forest industry to very old forests has been made possible by reclassifying a proportion of such terrain as 'negligibly disturbed forest'.[117] In Victoria in the 1990s—a state which contains some of largest and oldest hardwood trees on Earth—more than a quarter of a million hectares of forest 'lacking' sufficient dead branches in their 'upper stratum'—was placed outside the category 'old-growth'. This was an immensely important administrative moment because it fundamentally altered the extent of old-growth forest which needed to be protected from logging under emergent Commonwealth policies.[118] A key effect of this was to expose some of the most carbon dense forests to logging.[119] This occurred *not* by way of physical confrontation or industrial protest but through the application of a particular 'scientific' and legislative approach to (re)categorising forest growth stages. By a stroke of a pen, in other words, the forest industry maintained access to high quality marginally disturbed forest while appearing to have curtailed its long-term reliance on old-growth timber.

There are several implications for criminology which stem from the above. First, it is essential to engage in work which explicates the dark figure of carbon emissions, and for that matter, the dark figure of species decline, injury and loss. What Sutherland did for the hidden reality of white-collar crime, others need to do for these domains. And what Becker and Goffman did in relation to the (in)capacity of various agents to resist the category 'criminal' or 'deviant', others need to do for greenhouse gas producing sectors (forestry, mining, transport, electricity, and so on).

[116] See J Dargavel, *Fashioning Australia's Forests* (Melbourne, Oxford University Press, 1995).

[117] See Halsey, above n 107, 194–204.

[118] Ibid.

[119] Assuming a carbon density of at least 350 t/ha (the rate used previously for old-growth forest), this negligibly disturbed forest held 105Mt of carbon (preventing around 385Mt of CO_2 from entering the atmosphere). It is important to consider that this is the situation in just one forest management area (East Gippsland) within one state (Victoria) in Australia. The carbon densities of other areas have also been similarly recast.

This, quite clearly, is no easy task. The emergence of ecological problems on a world scale requires that criminologists commit themselves to interdisciplinarity in unlikely directions—not just to sociology, philosophy, economics, history, law or political science but to ecology, biology, chemistry, hydrology, physics, geography and the like. As a discipline, criminology needs to deconstruct the epistemologies and truth claims around which various governments and industries build their case for climate action, or, as the case may be, inaction. Interdisciplinary expertise and analysis is needed to determine when and how crime, harm or pollution are being defined down as environmentally benign (for example, clearfelling, uranium mining, oil refining) and, correlatively, when other types of behaviour are being 'defined up' as environmentally harmful (for example, picking wild flowers from a national park, driving a vehicle in a prohibited area, exceeding the recreational bag limit for fish). Third, and building upon recent work in the area of desistance from crime,[120] it is important to understand the conditions which prevent state, corporate or individual desistance from particular types of ecological harm—in this case, excessive GHG emissions. This would necessarily involve a detailed investigation of *persistent* emitters even if they are not labelled as such. Presently, there is a worrying dynamic in play concerning western nations and their claim to good environmental citizenship. From a desistance perspective, many countries seem all-too-eager to proclaim themselves and be viewed by other nations as ecologically responsible (as satisfying the conditions for 'secondary desistance')[121] without having demonstrably 'knifed-off' from their ecologically injurious pasts (without having achieved primary desistance). Finally, there is, I believe, an important methodological point to be gleaned from the foregoing analysis. Just as the lived experience of small groups of offenders have helped to problematise the validity of general ('catch-all') theories of crime, the events occurring in a single industry in one country can help problematise orthodox ways of framing international environmental regulations. This last point speaks to the need to undertake timely micro-political studies of how resource use at the local level (of forests, minerals, fossil fuels, and so forth) contributes to and might possibly mitigate climate change on a global scale.

[120] S Maruna, *Making Good: How Ex-Convicts Reform and Rebuild Their Lives* (Washington, DC, American Psychological Association, 2001).

[121] S Maruna, T LeBel, N Mitchel and M Naples, 'Pygmalion in the Reintegration Process: Desistance From Crime Through the Looking Glass' (2004) 10(3) *Psychology, Crime and Law* 271.

11

Personal Carbon Trading: Towards Sustainable Consumption in an Age of Climate Change and Energy Constraints

PETER DORAN

Choice is beneficial up to a point. But limitations, restrictions and boundaries can have a strangely liberating effect.[1]

PERSONAL CARBON TRADING COULD BRING ABOUT A STEP CHANGE IN UK SUSTAINABLE CONSUMPTION POLICY[2]

WITH THE GLOBAL carbon market and consumer lifestyles moving up the agenda of international climate change negotiations alongside growing concerns about 'peak oil' and energy security, a number of leading politicians, think tanks and parliamentary committees have turned their attention to proposals for a Personal Carbon Trading (PCT) scheme in the United Kingdom. Such an initiative would represent a decisive step change in the UK Government's approach to energy demand management and sustainable consumption. While the question of individual freedom and the important legitimating role played by consumption in liberal democracies have forced policy makers to tread lightly around the implementation of policies designed to promote more sustainable consumer choices, policy trends under the neoliberal order are also consistent with a partial devolution of climate security policy to the individual citizen and a concomitant renegotiation of the meaning of freedom when lifestyles are

[1] A Sigman, 'The Explosion of Choice: Tyranny or Freedom?', cited in the Report *I will* if *you will*: *Towards Sustainable Consumption* (London, Sustainable Development Commission and the National Consumer Council, 2006). www.aricsigman.com/research.html.

[2] An earlier version of this chapter appeared in D French (ed), *Global Justice and Sustainable Development* (The Netherlands, Martinus-Nijhoff, 2010).

placed on the negotiating table. A key characteristic of the contemporary world era is the co-evolution of manufactured risk and an expansion of individual decision-making responsibilities.

I want to suggest that Personal Carbon Trading may be consistent with the governmentality of advanced liberal democracies, wherein responsibilities for the management of risk are increasingly laid at the doorstep of individual citizens. Under conditions of ecological stress, citizens will increasingly be called upon by Government to navigate between their roles as consumers and calls for environmental citizenship.

Foreshadowed by climate change, and driven home by the global financial crisis, governance in the UK is now shaped by the imperatives of the 'risk society'[3] in a world in which the constitutive categories of inside and outside have permanently altered. National centres of power in the privileged islands of high modernity, where 'energy imperialism' has historically secured a precarious existence,[4] can no longer convincingly patrol and discipline the boundaries of their responsibilities. The ecological nature of contemporary risk induced by industrial technology has fallen out with the old categories of its over-productive system.

For Connolly,[5] late modernity is a time in which the worldwide web of systemic interdependencies has become more tightly drawn, while no political entity or alliance can attain the level of efficiency to master the system and its effects. There is a widening gap between the power of the most powerful states and the power they would require to be truly self-governing and self-determining.[6] No contemporary state can be inclusive enough to master the environment that conditions it; but the ideal of the late modern, democratic state, as the consummate agency of collective freedom, remains predicated on the assumption of this very capacity. The multilateral pursuit of climate change agreements, with their far-reaching implications for the design of our economies and the basis upon which access to environmental and atmospheric resources are shared (in space and across time), will demand that Governments re-visit the compact with their populations, notably in advanced liberal democracies where consumption and individual freedom are written into the fabric of democratic experience. In Connolly's words, as States re-negotiate the widening gap between their sufficiency—in other words the gap between world 'systematicity'[7] and state efficacy—states have begun to extend the operation of disciplinary power. The state is increasingly a medium through which world-systemic pressures—financial

[3] U Beck, *Risk Society: Towards a New Modernity* (London, Sage Publications, 1992).

[4] W Leiss, *The Limits to Satisfaction: On Needs and Commodities* (London, Marion Boyars, 1978).

[5] W Connolly, *Identity/Difference: Democratic Negotiations of Political Paradox* (London, Cornell University Press, 1991).

[6] Ibid 23–24.

[7] Refers to a putative series of psychological regularities.

and ecological—are transmitted to its domestic constituencies as imperatives of domestic discipline experienced as compulsions to choose and re-define a life course. It is a set of choices where the valorisation of the optional takes the place of the optimal.

Once the sovereign authors of designs on order and progress, late-modern states today find that the process is going into reverse: they must now mediate a more chaotic world of globalising risk. The on-going shift in geopolitical and financial power will frame the domestic experience of climate change politics and policies; as advanced liberal democratic states cede geopolitical and financial power to processes of globalisation (and an assertion of power by emerging economies), so the experience of disciplinary power by domestic constituencies of consumer-citizens will also register change, including a re-negotiation of their notions of freedom and the landscape of choice.

This chapter draws on Nikolas Rose's reading of Foucault for insights into the modern construction of 'citizen' and 'consumer', and, in particular, Rose's[8] discussion on the new specification of the subject of government to propose that the concept of Personal Carbon Trading sits at the apex of a number of debates of interest to the environmental policy community, including climate change, sustainable consumption and individual freedom in an era of global environmental and resource constraints.

Low Carbon Lifestyles Come to the Fore in National and International Policy Debates on Energy and Climate Change

Against the background of climate change negotiations and the mainstreaming of concerns about 'peak oil', energy consumption and sustainable lifestyles have come to the fore in both national and international policy debates. They were part of the subtext when the House of Commons Environmental Audit Committee issued a supportive report on Personal Carbon Trading (PCT) in May 2008,[9] and in the IPCC Chair, Dr Rajendra Pachauri's follow up to the Panel's Fourth Assessment Report's (2007) when he called in 2008[10] for a cut back in meat consumption as an individual contribution to curbing climate change.

More recently, the UK's All Party Parliamentary Committee on Peak Oil (APPCPO) commissioned a report in 2011 on Tradable Energy Quotas (TEQs),

[8] N Rose, *Powers of Freedom: Reframing Political Thought* (Cambridge, Cambridge University Press, 1999).

[9] *Fifth Report of Session 2007–08, House of Commons Environmental Audit Committee (HC 565)* Personal Carbon Trading (London, The Stationery Office Limited, 2008).

[10] *The Observer*, 'UN says eat less meat to curb global warming' (Sunday, 7 September, 2008) 1.

which sets out the case for a personal carbon trading role in addressing climate change and the depletion of fossil fuels in a way that can deliver a managed but steep reduction in the use of fossil fuels while forestalling fuel poverty by guaranteeing fair entitlements. It is within the context of the so-called 'triple crunch' of climate change, warnings of a looming energy crisis linked, in part, to 'peak oil', and the financial downturn that the language of 'carbon budgets', 'cap and trade' and 'carbon offsetting' has entered into policy discussions and even gained a slate of celebrity endorsements and champions, for better or for worse.[11] In the UK, any Personal Carbon Trading scheme would also have to operate in tandem with the European Union Emissions Trading System (ETS).[12] Noting the challenge of achieving an 80 per cent reduction in emissions by 2050 as set out in the UK Climate Change Act, with cuts of 34 per cent by 2020, Fleming and Chamberlin[13] argue that while there are over one hundred UK policies designed to impact on carbon emissions 'there is as yet no coherent structure in place to ensure that this emissions cap is respected'. While it is tempting to think of a tightening cap on emissions as a solution in itself, the true challenge as identified by Fleming and Chamberlin is to transform society so that it can thrive within this limit, if political pressure to loosen or abandon the cap altogether is not to overwhelm the ability of politicians to deliver a top-down approach.[14] It is this compelling rationale for Personal Carbon Trading that suggests a unique homology with the emerging debates on sustainable consumption and the need to transform lifestyles and economic signals in a way that shifts the debate from one of austerity to enabling modes of life and energy-use that can flourish within limits. Moreover, the IPCC evidence base that is helping to shape current UNFCCC negotiations has left policy-makers in no doubt that lifestyle change will have to be part of this mix when it comes to designing policies and measures that match the ambition of the anticipated emissions reduction and mitigation targets and goals beyond the final Kyoto Protocol budget period of 2008–2012. In both the transport and household sectors, the main focus of UK policies has been on uncovering technical efficiencies to bring about energy savings, largely successfully in the case of the household sector and less successfully in the case of the transport sector. Kerr and Battye[15] have noted, however,

[11] Alex James, of the UK 'iconic band' Blur appeared on the BBC's prime-time Sunday *Politics Show* in November 2006 to discuss personal carbon trading, based on an allocation of 'carbon credits' to help individuals monitor and contain their annual carbon emissions.

[12] A Kerr and W Battye, *Personal Carbon Trading: Economic Efficiency and Interaction With Other Policies* (London, Royal Society of Arts (RSA), 2008). www.theRSA.org/carbon.

[13] D Fleming and S Chamberlin, *Tradable Energy Quotas: A Policy Framework for Peak Oil and Climate Change, House of Commons All Party Parliamentary Group on Peak Oil* (London, The Lean Economy Connection, 2011).

[14] Ibid, 40.

[15] Ibid.

that the potential 'space' in the UK policy framework for a PCT scheme to deliver emission reductions will be most widely available in those areas where Government has, until now, shied away from policies designed to induce a change in consumer behaviour. For example, in changing demand for transport services and controlling household energy demand.

The Government reticence that has characterised its response to the concept of Personal Carbon Trading to date is linked to a long-standing nervousness about consumer reaction to a more interventionist approach to shaping consumer behaviour, or a move to a sufficiency approach. For Sachs,[16] Princen[17] and others, the efficiency perspective that dominates current discussion on the 'sustainable production' side of the production and consumption equation must be accompanied by a *sufficiency* debate on the consumption side. The transition towards sustainability can only be achieved through a twin-track approach, which brings about an intelligent reinvention of the means as well as a prudent moderation of ends. The questions of 'ends' and 'what the economy is for?' draw policy makers into uncharted territory in the consumption debate. Is it possible, for example, to imagine *flourishing* communities where an appreciation of limits and self-restraint has been deeply embedded or re-covered in our culture and society? Few governments in advanced liberal democracies have begun to pursue the logic of the sufficiency debate, and for understandable reasons.

Part of the difficulty with the emerging governmental challenge of redefining wealth and prosperity, is that it runs against the grain of embedded assumptions about consumption in liberal democracies. As Offer[18] reminds us, modern consumption theory assumes that rational consumers make choices that are well informed, far-sighted, and prudent. Consumers reveal their preferences by means of market choices, and market choices correspond to their well-being ('welfare'). Taking account of the expected value of lifetime wealth, consumers maximise welfare by smoothing consumption over the life cycle. An 'invisible hand' then acts to aggregate individual choices to maximise the economic welfare of society. Offer[19] points out that a great deal is at stake in the model of 'revealed preferences' as the source of well-being. Indeed, it may be nothing less than the conceptual underpinning of liberal society. The doctrine regards the free exercise of market choice as not only economically efficient, but also as a vital human aspiration. It is, in part, for this reason that political leaders have been slow to pursue the 'sustainable consumption' agenda since the 1992 UN Conference on

[16] W Sachs, *Planet Dialectics: Explorations in Environment and Development* (London, Zed Books, 1999).

[17] T Princen, M Maniates and K Conca (eds), *Confronting Consumption* (Cambridge, MA, MIT Press, 2002).

[18] A Offer, *The Challenge of Affluence: Self-Control and Well-Being in the United States and Britain Since 1950* (Oxford, Oxford University Press, 2006).

[19] Ibid.

Environment and Development. The introduction of a PCT system in the UK would represent a radical advance on the public discourse and implementation of sustainable consumption in the context of energy savings and the transition to a low carbon economy. It would entail a level of intervention in shaping consumer demand beyond anything implemented, to date, in pursuit of consumer behaviour change (alongside supply side measures), requiring a shift that would inevitably provoke far-reaching questions about the role of the state and the freedom of the individual citizen-consumer.

If future international commitments under the Kyoto Protocol are to match the demands of the current IPCC science, climate change is likely to force the hand of policy makers who have adopted a tentative approach to interventions in consumer behaviour. Specifically, if current climate change policies manifestly fail to make a dent in emissions from the personal sector (that account for some 40% of UK emissions), PCT may no longer be regarded as an idea before its time[20]. The Intergovernmental Panel on Climate Change has made it clear that changes in lifestyle and behaviour can contribute to climate change mitigation across all sectors. In particular, the IPCC Fourth Assessment Report calls for changes in lifestyle and consumption that emphasise resource conservation and contribute to the development of a low carbon economy that is both sustainable and equitable.[21] The first real world experiment in Personal Carbon Trading, in Norfolk Island, a self-governing Australian protectorate off New South Wales, with a resident population of 1,750, will attempt to link individual carbon emissions reductions with improvements in health. With the agreement of the Norfolk Island legislature, three Australian universities will conduct the study, including:

a. A pilot test of the effectiveness of a PCT system in reducing carbon emissions and improving chronic disease risk (particularly obesity);
b. To assess the individual and likely political acceptance of such a scheme for use in a whole country environment; and
c. To develop and pilot test shareholder involvement of the community in the scheme.

The experiment has been designed as a 'stealth intervention' that will have goals that extend beyond the reduction of carbon emissions. It is anticipated that the intervention will impact on lifestyles in a way that will also help reduce obesity and chronic disease, by increasing personal energy use,

[20] Institute for Public Policy Research (advance draft), *Plan B?: The Prospects for Personal Carbon Trading. A Report by Jenny Bird and Matthew Lockwood* (London, IPPR, 2009).

[21] IPCC, 'Fourth Assessment Report of the Intergovernmental Panel on Climate Change (AR4)' 2007(IPCC). www.ipcc.ch/ 16.

changing attitudes to 'superfluous' consumption and travel, and influencing infrastructure design.[22]

In the decade since Agenda 21 (UNCED 1992),[23] technological development and innovation have increased resource efficiency at some levels and in some sectors. However, these developments have not amounted to an adequate response to address critical patterns of unsustainable consumption and production. At the World Summit on Sustainable Development (WSSD) in 2002, negotiators revisited the sustainable consumption agenda and agreed the Johannesburg Plan of Implementation (Chapter III) which includes a call for the development of a 10-year framework of programmes[24] in support of regional and national initiatives to accelerate the shift towards sustainable consumption and production (SCP), with a focus on policy responses in the EU and OECD countries. The 10-year framework[25] is being developed to strengthen international co-operation and increase exchange of information and best practices to facilitate the implementation of national and regional programmes to promote sustainable consumption and production. In March 2003, the European Council (the EU Heads of State or Government) identified sustainable consumption and production as one of the key priorities of the EU in its follow up to the WSSD.

Seyfang[26] has shown how tensions and competing perspectives have been part and parcel of the sustainable consumption concept since its introduction in Agenda 21 at UNCED in 1992, the first global conference on the environment to implicate over-consumption in the developed world as a direct contributor to unsustainability. The proposed solutions included promoting eco-efficiency and using market instruments for shifting consumption patterns; and a second recommended that governments should develop 'new concepts of wealth and prosperity which allow higher standards of living through changed lifestyles and are less dependent on the Earth's finite resources and more in harmony with the Earth's carrying capacity'.[27]

The definition of sustainable consumption narrowed post-1992 as it 'evolved' in the course of negotiations at a range of international policy

[22] Egger, G, *Investor/Sponsor Information: A Best Practice Trial to Assess the Effects of Personal Carbon Trading (PCT) on Carbon and Chronic Disease Reduction.* The NICHE (Norfolk Island Carbon/Health Evaluation) Study (2010). See also Egger, G, 'Personal Carbon Trading: A Potential 'Stealth Intervention' for Obesity Reduction?' (2007) *Medical Journal of Australia* 185–87.

[23] United Nations Conference on Environment and Development, Rio de Janeiro, 3–14 June (1992).

[24] Report of the World Summit on Sustainable Development Johannesburg, South Africa (26 August—4 September 2002) A/CONF.199/20/Corr.1.

[25] The 'Marrakech Process' was launched at the first international expert meeting on the 10-year framework held in Marrakech, Morocco, 16–19 June 2003, organised by DESA's Division for Sustainable Development and UNEP.

[26] G Seyfang, 'Shopping for Sustainability: Can Sustainable Consumption Promote Ecological Citizenship' (2005) 14 (2) *Environmental Politics* 290–306.

[27] Agenda 21, UNCED 1992: s 4.11.

arenas and became more widely accepted as a policy goal. Seyfang[28] cites, for example, the work of the OECD during the 1990s and its conclusion that market failure was the prime cause of unsustainability. She concludes that, within the strong market-liberal perspective, as reflected in OECD analyses, governments are encouraged to correct prices and provide regulatory frameworks to influence producers and stimulate eco-efficiency, and offer consumers more green choices. She believes that an emphasis on decoupling economic growth from environmental degradation, to be achieved through a range of market-based measures and education initiatives, represents a narrowing of the sustainable consumption agenda from the initial possibilities of redefining prosperity and wealth and radically transforming lifestyles. The focus, instead, is now on improving resource productivity and marketing 'green' or 'ethical' products.

A STEP IN THE RIGHT DIRECTION: FLOURISHING

In contrast, the UK Sustainable Development Commission (SDC), in one of its final and most influential publications before it was stood down by the incoming Conservative-Liberal Democratic Government, placed a question mark over the 'social logic of consumerism' in its report, *Prosperity Without Growth: the transition to a sustainable economy*.[29] The author, Tim Jackson, argues that prosperity is not synonymous with material wealth. Rather:

> ... prosperity has to do with our ability to flourish: physically, psychologically and socially. Beyond sheer subsistence or survival, prosperity hangs on our ability to participate meaningfully in the life of society.[30]

Drawing on the work of Sen,[31] Jackson notes that while the material requirements for physiological flourishing and for a 'life without shame'[32] tend to be fairly similar across societies, it requires a *more* expensive bundle of goods and services in a society that is generally richer and in which most people already have, say, means of transport, affluent clothing, radios or

[28] See Seyfang, above n 19, 290–306.

[29] T Jackson, *Prosperity Without Growth: the Transition to a Sustainable Economy* (London, Sustainable Development Commission, 2009) 86.

[30] Ibid.

[31] A Sen, *The Living Standard* (Oxford, Economics Papers, 1984).

[32] This is a reference to the work of Adam Smith and his insight on the importance of shame in social life. In the Wealth of Nations, Smith cites the example of the linen shirt. While the shirt could not be deemed a necessity of life, Smith noted, that in 'present times', through the greater part of Europe, a creditable day labourer would be ashamed to appear in public without a linen shirt, 'the want of which would be supposed to denote that disgraceful degree of poverty which, it is presumed, no body can well fall into without extreme bad conduct': A Smith, *An Inquiry Into the Nature and Causes of the Wealth of Nations* (New York, Modern Library, 1937, reprint) 821.

television sets, and so on. The same absolute level of capabilities may have a greater relative need for incomes (and commodities). Jackson's striking conclusion is that, if we take for granted the importance of material commodities for social functioning, there is never any point at which we will be able to claim that enough is enough. The baseline for social functioning is always the *current level of commodities*. And the avoidance of shame—a key feature of social flourishing—will drive material demand forward relentlessly:

> Social progress appears to rest on a 'social trap' wherein social progress depends on the self-reinforcing cycle of novelty and anxiety. Material throughput will inevitably grow. And the prospects for flourishing within ecologic limits evaporate. The greatest threat to long-term prosperity is the continuing surge of materialism, and from an economic model that perpetuates it.[33]

Noting evidence of a 'social recession' across the political spectrum in the UK (including rising rates of anxiety, clinical depression, addiction, low morale at work, etc), Jackson also describes Soper's[34] observation of a growing appetite for 'alternative hedonism' or sources of satisfaction outside the conventional market. Tim Kasser's[35] work on values and prosperity seems to lend some empirical support to Soper's observation. Empirical research conducted by Kasser and others[36] on people's values and goals appears to show that the set of values that might act as a counterweight to the materialistic values so dominant in a profit-driven, consumer society are the same set that are associated with high levels of personal well being, and the same set of values that are conducive to civil social behaviour. Kasser notes:

> ... goals for financial success, image, and popularity tend to stand in opposition to three other 'intrinsic' goals: Self-acceptance (which involves trying to grow as a person), affiliation (which involves having close, intimate relationships with family and friends), and community feeling (which involves trying to help the broader world be a better place). What this finding means is that the more people focus on materialistic aims for money, image, and status, the more difficult it is for them to also focus on growing as a person, intimately connecting with other people, and contributing to the world at large.[37]

[33] Jackson, above n 29, 88.

[34] K Soper, *Exploring the Relationship Between Growth and Wellbeing* (London, Sustainable Development Commission, 2008), www.sd-commission.org.uk/pages/redefining-prosperity.html.

[35] T Kasser, *A Vision of Prosperity* (London, Sustainable Development Commission, 2007), www.sd-commission.org.uk/pages/redefining-prosperity.html.

[36] K Brown and T Kasser, 'Are Psychological and Ecological Well-Being Compatible?' (2005) 74 *Social Indicators Research* (2005) 349–68.

[37] Kasser, above n 29, 3.

Most interestingly, for Kasser,[38] is the finding that the intrinsic aims for self-acceptance, affiliation and community feeling are also the values and goals that empirical research demonstrates promote personal happiness, positive social involvement, and ecologically sustainable behaviour. According to Kasser,[39] studies show that happier people place a relatively high emphasis on the goals of self-acceptance, affiliation and community feeling. By caring about such aims, the research suggests that people become increasingly likely to create lifestyles for themselves that satisfy the psychological needs required for psychological thriving. The pursuit of these intrinsic goals has also been associated with caring more about ecological sustainability and being less greedy with limited resources.[40]

Kasser has begun to provide empirical evidence to support the view that it is not only possible for people to live happier lives while exercising higher levels of environmental responsibility, but that people have already embarked on such life paths. Jackson[41] suggests that the findings are extraordinary because they suggest that there is a double or triple dividend in a less materialistic life: people are no unhappier and live more sustainably when they favour intrinsic goals that embed them in family and community. Flourishing within limits is a real possibility.

It is often reported that a gap shows up in research into consumers' stated preferences and their actions. In the words of the American activist, Saul Alinsky,[42] most people today are eagerly groping for some medium, some way in which they can bridge the gap between their morals and their practices. Economic instruments in pursuit of relatively popular and widely understood governmental objectives, such as environmental sustainability, do not merely impose constraints; for the majority of a population, the instrument acts as a prompt or a cue to pursue already embedded value judgments. It is in this sense that a system such as Personal Carbon Trading (PCT), while perceived as a potential source of limitations, restrictions and boundaries, could have a strangely liberating effect on consumer-citizens who have grown increasingly concerned at the prospect of catastrophic climate change on their lives, the lives of their children, and their wider communities of fate. Moreover, as Fleming and Chamberlin have argued, a system of Personal Carbon Trading represents a shift away from an *extrinsic* approach to incentivising shifts in behaviour (eg using tax incentives) and a shift towards an *intrinsic* mechanism that provides reasons to want to reduce dependency on fossil fuels, to plan ahead, and to co-operate

[38] Ibid.
[39] Ibid.
[40] Brown and Kasser, above n 30.
[41] Jackson, above n 29, 88.
[42] S Alinsky, *Reveille for Radicals* (New York, Vintage Books, 1969) 94.

with others in applying ingenuity.[43] The concept is based on 'pull' or on principles of lean thinking developed in Japan during the post-war period, which recognises that the people best placed to understand and cope with complex local tasks are those who are faced with the immediate challenge of completing those tasks on the ground.

PERSONAL CARBON TRADING

A system of 'Domestic Tradable Quotas' (DTQs) was first proposed by Dr David Fleming, a London-based policy analyst, in 1996.[44] Since July 2003, a project at the University of Manchester, funded by the Tyndall Centre for Climate Change Research, has been assessing the feasibility of DTQs as an instrument of public policy.

In brief, the DTQ model is a 'cap and trade' scheme that would operate in three phases:

a. Setting the carbon budget: establishing the maximum quantity of greenhouse gases that may be emitted from energy use in a given year by a nation implementing the DTQs scheme. Under the scheme, the budget is reduced year-on-year in line with national emissions reductions targets.
b. Surrendering carbon units: each carbon budget is divided into carbon units, with one carbon unit representing 1 kg of carbon dioxide equivalent;
c. Acquiring units for surrender: Carbon units are allocated to eligible individuals by government on a free and equal per capita basis. The proportion of total carbon units allocated to individuals under the Entitlement is equal to the proportion of total energy emissions arising from individuals' purchase of fuel and electricity over a given period prior to the introduction of a DTQs scheme (eg in the UK this is about 40%).

Emission rights are allocated to and surrendered by end-purchasers of fuel and electricity. These rights are allocated to individuals free and on an equal per capita basis while organisations would purchase the units they require on a national carbon market. Whenever individuals and organisations purchase fuel or electricity, they would be required to surrender to the retailer carbon units to cover the quantity of fuel or electricity purchased. Energy efficient individuals who surrender less emissions rights than they are allocated could sell their surplus onto the national market and individuals requiring additional rights or credits to those they have been allocated would have to purchase these on the market.

43 Above n 13. 23.
44 See www.dtqs.org and www.teqs.net.

With carbon dioxide emissions from individuals on the rise and representing a significant 42 per cent share of the UK's emissions, changes in both collective and *individual* behaviour have taken on a new urgency. The two main emitting activities in the UK household are space heating within the home and personal car travel. Within the home, about 75 per cent of emissions relate to provision of space and hot water heating, with the remainder relating to cooking, appliances and lighting. Holiday air travel comprises nearly 28 per cent of total travel-associated emissions.[45]

Fleming's original model for personal carbon trading, Tradable Energy Quotas, was first launched in a 1996 magazine article, in *Country Life*. This was followed by a discussion paper and a series of policy discussions involving, among others, the UK Department of Trade and Industry, the European Commission, and various seminars sponsored by Members of the House of Commons. Think tanks took up the concept during this period, including work by the Environmental Change Institute (Oxford), the National Economic Research Council, the Royal Society of Arts, and the Institute of Public Policy Research.

The concept of PCT took a significant step out from the shadow lands of think tanks and party research briefings when the former Labour Environment Minister, David Miliband MP, included favourable comments in a speech delivered in 2006. The original UK Climate Change Bill[46] published in November 2007 included provision for the introduction of domestic emissions trading regimes, holding open the possibility of secondary legislation for the introduction of a personal carbon trading system at some time in the future. A number of projects and studies brought new prominence to the concept, including:

— The Royal Society of Arts (RSA) launched a 3 year policy development project exploring ways to involve individuals in addressing climate change.
— An RSA-commissioned study (September 2007) recommended that personal carbon trading would be feasible by 2013 and could play a role in re-establishing trust between citizens and government on climate change.[47]
— The Centre for Sustainable Energy published a Defra (Department of the Environment, Food and Rural Affairs) commissioned study (November 2006) on personal carbon trading and its variations.[48]

[45] A Kerr and W Battye, *Carbon Limited: Exploring Personal Carbon Trading* (London, Royal Society of Arts, 2008). www.theRSA.org/carbon.

[46] The UK Climate Change Act was enacted by Royal Assent 26 November 2008

[47] Royal Society of Arts, *Personal Carbon Trading—the Idea, Its Development and Design* (London, Royal Society of Arts, 2007), www.rsacarbonlimited.org/viewarticle.aspa?pageid=797.

[48] S Roberts and J Thumim, *A Rough Guide to Individual Carbon Trading: the Ideas, the Issues and the Next Steps* (London, Centre for Sustainable Energy/DEFRA, 2006).

Research commissioned by Defra in August 2006 and conducted by the Centre for Sustainable Energy[49] examined the advantages and disadvantages of different approaches and concluded that a personal carbon allowance and trading system had the potential, with further research, to achieve emissions savings in a fairer way than carbon taxes. The Environmental Audit Committee[50] came down in favour of the PCT as the kind of measure that might be needed to bring about behavioural change, given that existing initiatives are unlikely to being about behavioural change on the scale required, 'with many individuals choosing to disregard the connection between their own emissions and the larger challenge.'[51] Their conclusion was supported by findings in the work of Roberts and Thumim[52] that the potential of personal carbon trading lay in its ability to deliver guaranteed reductions in emissions from individuals, a 'theoretical certainty' that was not shared by all instruments. Personal carbon trading could guarantee a reduction in emissions because it places a ceiling on the carbon available for consumption, rather than seeking solely to reduce demand. In this sense, PCT systems represent the holy grail of sustainable consumption, which has been limited—insofar as implementation is concerned—to the committed niche groups of ethical consumers.

NAVIGATING RISK AND CLIMATE (IN) SECURITY

Personal Carbon Trading would represent a step change, or much more interventionist shift, in governmental approaches to sustainable consumption. The ambition of UK Government policy in the context of climate change—given the risks now associated with inaction—may provide the conditions for a new approach to shaping consumer choices in the energy field.

One of the features of globalised risk production is the all-too-apparent limits of an individual State when it comes to delivering on the traditional guarantees of security and welfare. In an age of contingency—and exposures to insecurity—the State manages and partially devolves that experience of contingency together with some of the responsibility. As the UK confronts choices in the face of the exemplary conditions of global contingency signalled by climate change, any new articulation of the relationship between the State and Citizen in meeting the challenge will also be framed by an established neoliberal articulation of the individual's share in the navigation of risk.

[49] S Roberts and J Thumim, *A Rough Guide to Individual Carbon Trading: the Idea* (London, Centre for Sustainable Energy, 2006).

[50] House of Commons Environmental Audit Committee, above n 45, 8.

[51] Ibid 9.

[52] Roberts and Thumim (2006), above n 48.

To demonstrate the way in which the global security complex of energy and climate change calculations is closing in on individual choice and behaviour, the UK's first 'National Security Strategy'[53] is a useful starting point. The document is reflective of the transformation in the discourse of security that has taken place under the signs of the emergence of a multi-polar world, 'international terrorism', weapons of mass destruction, pandemics and transnational crime. In turn, the report states, 'These and other threats and risks are driven by a diverse and interconnected set of underlying factors, including climate change [and] competition for energy.'[54]

Inter-state and inter-regional factors are now joined by newly acknowledged 'threats and risks' and responses that reflect the intimate nature of distance in an era of globalisation in which the relationship between the local and global has suffered a loss of innocence. Environmental security will reach all the way down to future calculations around the consumption decisions of the householder. The Security Strategy states that climate change is potentially the 'greatest challenge to global stability and security' alongside energy security, calling for an urgent, integrated and international response:

We are working at all levels—from our role in the international community and the EU, to national level, local authorities and communities, and in partnership with businesses and citizens—to make the technological and behavioural transition to a low-carbon economy.[55]

As Bislev, Salskov-Iversen and Krause Hansen[56] have noted, the maintenance of national and international social order as well as the management of the physical environment are now presented as crucial tasks for the reproduction of security. The writers are interested in the introduction of Public Management techniques as part of the transition in the governance role of the state. This is a trend that has brought market mechanisms and managerial techniques into spheres of society not hitherto permeated by the market. These include such areas as the personal sphere, the family, voluntary sectors and the state. In sum:

> The state is now represented increasingly as an institution doing two things: delivering collective services to a large and diversified group of customers, and inspiring groups and individuals to govern themselves in certain directions.[57]

All of this has implied the implementation of new forms of regulation (in the security and other fields), often based on marketisation and consumerism. Two technologies of government are at work here. First, governing through the installing of 'technologies of the self', working through

[53] Cabinet Office, *National Security Strategy* (Cabinet Office/HMSO, 2008) 50.

[54] Ibid 3.

[55] Ibid 50.

[56] S Bislev, D Salskov-Iversen and H Krause Hansen, 'Globalization, Governance and Security Management', Working Paper No 43, Presented at the American Political Science Association Annual Meeting, San Francisco, August 29–September 1 2001.

[57] Ibid 5.

a process where discourses are developed for self-governing. Groups, organisations and individuals will employ the ideas, understandings and forms of knowledge contained in these discourses. Secondly, the authors detect a tendency towards the state becoming preoccupied with governing *as such* or reflexive government.[58]

Both dimensions are at work in the decision by the UK Government to publish its first *National Security Strategy*.[59] There is an attempt by the Government to raise and re-frame security in the context of new threats and new expectations about the respective roles of stakeholders, with the Government now positioned as one, albeit primary actor, among the many, including the individual citizen.

Lemke on Governmentality and Neo-liberalism

At the heart of this question, and reflective of a larger question about the relationship between the economy and politics, is the new dominance of economic actors in both perpetuating the notion of an autonomous 'economic' sphere and its counterpart in the consumer-citizen. As noted by Lemke,[60] a key feature of neo-liberal rationality is the congruence it endeavours to achieve between a responsible and moral individual and the role of an economic-rational individual. It aspires to construct responsible subjects whose moral quality is based on the fact that they rationally assess the costs and benefits of a certain act as opposed to other alternative acts.

As the choice of options for action is, or so the neo-liberal notion of rationality would have it, the expression of free will on the basis of a self-determined decision, the consequences of the action are borne by the subject alone, who is also solely responsible for them. This strategy can be deployed in all sorts of areas and leads to areas of social responsibility becoming a matter of personal provisions.[61] In the context of consumerism, one expression of this shift from collective or 'social' to individualised responses to risk, including an expansion of consumer responses to threat, has been documented by Szasz.[62] Whereas consumption can be described as primarily a trait and occupation of individual human beings, consumerism is an attribute of a society[63] and acts to channel individualised responses to risk where the public realm has either retreated or is deemed to have failed

[58] Ibid 6.

[59] National Security Strategy of the United Kingdom: Security in an Interdependent World (London, Cabinet Office, 2008).

[60] T Lemke, 'Foucault, Governmentality, and Critique'. A paper presented at the *Rethinking Marxism Conference*, University of Amherst (MA) (September 21–24 2000) 12.

[61] Ibid 12.

[62] A Szasz, *Shopping Our Way to Safety: How We Changed From Protecting the Environment to Protecting Ourselves* (Minneapolis, University of Minnesota Press, 2007) 12.

[63] Z Bauman, *Consuming Life* (Cambridge, Polity Press, 2007) 28.

to protect the public interest. Bauman has argued that 'consumerism' is to be distinguished from consumption and is a type of social arrangement that results from recycling mundane, permanent and so to speak 'regime neutral' human wants, desires and longings into the principal propelling and operating force of society, a force that coordinates systemic reproduction, social integration, social stratification and the formation of human individuals, as well as playing a major role in the processes of individual and group self-identification and in the selection and pursuit of individual life policies. 'Consumerism', he believes, arrives when consumption takes over that linchpin role that was played by work in the society of producers.

The upshot, for Szasz, is what he describes as a form of 'inverted quarantine':

> ... the kind of response to threat that I am interested in is individualistic in both goal and method. A person who, say, drinks bottled water or uses natural deodorant or buys only clothing made of natural fibre is not trying to change anything. All they are doing is trying to barricade themselves, individually, from toxic threat, trying to shield themselves from it. Act jointly with others? Try to change things? Make history? No, no. I'll deal with it individually. I'll just shop my way out of trouble.[64]

Rather than working together with others [citizens] to get the political system to acknowledge and deal with problems, and acting in the modality of a 'citizen in a democratic society', a person who buys some product in order to shield herself from harm is acting in the modality of 'consumer'. Again, we can appeal to the work of Foucault to help explain this inevitable consumer-driven response to collectivised risk and the consumption of security. What we are witnessing is further evidence of a strategy of rendering individual subjects 'responsible', entailing a shift in responsibility for social risks and for life in society into the domain for which the individual is responsible (the domain of consumption) and transforming this into a problem of 'self care'.

GOVERNMENTALITY: THE CITIZEN MUTATES INTO A CONSUMER

Since the end of the 1980s, in advanced liberal democracies, government responses to neo-liberalism have even more firmly embedded the role of 'consumer' as a client of governmental services. We will now examine this transformation of governance and its implications for citizens and consumers.

The prospect of PCT in the UK invites a re-reading of a number of concepts introduced by Michel Foucault and developed by others, including

[64] Szasz, above n 62, 12.

Nikolas Rose[65] and Thomas Lemke.[66] As Clarke[67] has noted, through the increasingly influential concept of governmentality, post-Foucauldian work has enriched our understanding of the range of 'conduct' and its government beyond the politics-economics couplet of political economy, and challenged sociological conceptions of society and its constitution. On the decline of 'the social', Rose[68] observes, for example, that while social government has been failing since its inception, the solution proposed is no longer the re-invention of 'the social'. As 'society' dissociates into a variety of ethical and cultural communities with incompatible allegiances and incommensurable obligations, 'a new set of political rationalities, governmental technologies and opportunities for contestation begin to take shape'.[69] Under the new formula of rule associated with advanced liberalism, Rose[70] notes that rule seeks to de-governmentalise the State and to de-statise practices of government and detach the substantive authority of expertise from the apparatuses of political rule. Experts are relocated within a market governed by rationalities of competition, accountability and consumer demand:

> It [rules] does not seek to govern through 'society', but through the regulated choices of individual citizens, now construed as subjects of choices and aspirations to self-actualization and self-fulfilment.[71]

Revisiting Rose's[72] summary of the debates and attacks on 'social government'—in the current context of consumption, and consumerism—it is instructive to note the role attributed to the mass media and marketing strategies deployed in commodity advertising. Charting the collapse of the 'empire of social expertise', Rose notes how clients of expertise came to understand and relate to themselves and their 'welfare' in new ways. In a whole range of sectors, individuals came to reconceptualise themselves in terms of their own will to be healthy, to enjoy a maximised normality:

Surrounded by images of health and happiness in the mass media and in the marketing strategies deployed in commodity advertising and consumption regimes, narrativising their dissatisfactions in the potent language of rights, they organised themselves into their own associations, contesting the powers of expertise, protesting against relations that now

[65] N Rose, *Inventing Ourselves: Psychology, Power and Personhood* (Cambridge, Cambridge University Press, 1996); and N Rose, 'Governing 'Advanced' Liberal Democracies' in A Barry, T Osborne and N Rose (eds), *Foucault and Political Reason: Liberalism, Neoliberalism and Rationalities of Government* (Chicago, University of Chicago Press, 1996).

[66] Lemke, above n 60.

[67] J Clarke, 'Consumerism and the Remarking of State-Citizen Relationships', Paper prepared for ESPAnet Conference Oxford (2004) 9–11.

[68] N Rose, *Powers of Freedom: Reframing Political Thought* (Cambridge, Cambridge University Press, 1999).

[69] Ibid 135–36.

[70] Ibid 41.

[71] Ibid 41.

[72] Ibid 52.

appeared patronising and demeaning of their autonomy… In the face of the simultaneous proliferation, fragmentation, contestation and de-legitimation of the place of experts in the devices of social government, a new formula for the relation between government, expertise and subjectivity would take shape.[73]

The new strategies and rationalities of government that can be detected in 'advanced liberal' societies seek techniques of government that create a distance between the decisions of formal political institutions and other social actors, conceive of these actors in new ways as subjects of responsibility, autonomy and choice, and seek to act upon them through shaping and utilising their freedom. In the context of climate change, as Governments are increasingly forced to qualify the realm of freedom of choice, this pattern of government distancing and devolving responsibility will become a more acute imperative in the field of consumption as governments seek to protect or shield themselves from any perceived threat to their legitimacy founded on consumer expectations as defined since the post-Second World War era of mass consumerism.

Of particular interest here is Rose's[74] discussion on the 'new specification of the subject of government', which offers a number of clues about the designations of 'citizen' and 'consumer' in an advanced liberal democracy such as the UK. Rose notes that an enhancement of the powers of the 'client as consumer'—(consumer of health services, of education, of training, of transport, etc) has specified the subjects of rule in a new way as active individuals seeking to 'enterprise themselves', to maximise their quality of life through acts of choice, according their life a meaning and value to the extent that it can be rationalised as the outcome of choices made or choices to be made.

Miller and Rose,[75] writing on the 'affiliated' members of society—those who are considered 'included' by virtue of their access to the financial, educational and moral means to 'pass' in their role as active citizens in responsible communities—describe how individuals must now 'enterprise' their lives through active choice, within authoritative terms and limits that have become integrated within all the practices of everyday life, sustained by a heterogeneous array of 'civilised' images and devices for lifestyle promotion, including through the activity of 'ceaseless consumption'. They note that today, a strategic shift is occurring in the politics of security:

> Individuals are, once again, being urged by politicians and others to take upon themselves the responsibility for their own security and that of their families: to insure against the costs of ill health through private medical insurance, to make

[73] Ibid 52.
[74] Ibid 42.
[75] P Miller and N Rose, *Governing the Present* (Cambridge, Polity, 2008).

> provision for their future through private pensions, to take an active role in securing themselves against all that could possibly threaten the security of their chosen style of life.[76]

The 'new prudentialism'[77] uses the technologies of consumption—advertising, market research, niche marketing and so forth—to exacerbate anxieties about one's own future and that of loved ones, to encourage action to subdue these risks and negotiate fate by purchasing insurance. The ethics of lifestyle maximisation, coupled with a logic in which someone must be held to blame for any event that threatens an individual's quality of life, generates a relentless imperative of risk management not simply in relation to contracting for insurance, but also through daily lifestyle management.[78] These technologies do not have their origin or principle of intelligibility in 'the State'. Nonetheless, they have made it possible to govern in an 'advanced liberal' way. They have provided a plethora of indirect mechanisms that can translate the goals of political, social and economic authorities into the choices and commitments of individuals, locating them into actual or virtual networks of identification through which they may be governed. These include the extension of privatised responsibilities or 'risk management' in areas such as diet control and home security; while social work has given way to the private counsellor, self-help and expert phonelines. The regulation of conduct becomes a matter of each individual's desire to govern their own conduct freely in the service of the maximisation of a version of their own happiness and fulfilment that they take to be their own, but which nevertheless entails a relation to authority:

Individuals are to become 'experts of themselves', to adopt an education and knowledgeable relation of self-care in respect of their bodies, their minds, their forms of conduct and that of the members of their families.[79]

This is no zero-sum game. While there has been a downside to the new neoliberal designations, the new dispensation also holds out a new possibility of agency ... even 'ethical reconstruction as active citizens.'[80]

CONCLUSION: AN UNCERTAIN CLIMATE OF FREEDOM

The challenge of consumerism and sustainable consumption in the context of climate change is inherently a challenge to contemporary understandings of individual freedom. That is the significance of the debate on policies such

[76] Ibid 99–100.

[77] P O'Malley, 'Risk, Power and Crime Prevention' (1992) 21 (3) *Economy and Society* 252–75.

[78] Ibid.

[79] N Rose, *Inventing Ourselves: Psychology, Power and Personhood* (Cambridge, Cambridge University Press, 1996) 59.

[80] Ibid 60.

as Personal Carbon Trading. Late-modern states, including the UK, are confronting a moment of transition that will be characterised by a departure from 'secular assurances'[81], which have often been worked out through acts of consumption, to a reworking not only of the relationship between the citizen and the state, but a revisiting of our understandings of freedom and felicity, and the conditions that made possible our current understandings of reason and subjectivity,[82] which have been deeply informed by the economic, institutional and intellectual technologies of world society.

Roy[83] traces one critical transformation back to Thomas Hobbes' fateful rejection of the notion of ultimate purpose and the reduction of our understanding of 'felicity' to 'a continual progress of the desire from one object to another; the attaining of the former being still but the way to the latter'[84]. Roy[85] concludes that with this revision, our needs became endless in two senses: on the one hand, needs came to be detached from higher life purpose and, on the other, they began to expand and proliferate endlessly. The endlessness of life came to be grounded in the acquisition and accumulation of material resources such as wealth, prestige and power. 'Commodious living', in the Hobbesian sense, came to constitute the prime motivating force of life in modern times. Friedrich Hölderlin[86] observed that this shift from simple to commodious living required a condition of the highest cultivation that could establish accord within and among men through self-discipline. In meeting the demands of his or her nature, the individual must—in order to protect collective well-being from possible harm—exercise a higher control over the self. Instead, advances in science and technology, and apparent control over nature, have been accompanied by a culture of consumerism that has helped to erode auto control and further conflate the goods life with any sense of the good life.

One of the most intriguing questions the modern citizen faces in this age of limits—an age in which it appears that the anticipated exhaustion of resources and pollution sinks is matched by the psychic exhaustion of a once universal modern imaginary that has culminated in an age of sovereign consumers demanding the best of all possible worlds—is the ageless question of freedom. On the meaning of freedom signalled by the rise of the ecological movement, Eckersley[87] once suggested that the new project entails much more than a simple reassertion of the modern emancipatory ideal

[81] Connolly, above n 5.

[82] H Caygill, 'Violence, Civility and the Predicaments of Philosophy' in D Campbell and M Dillon (eds), *The Political Subject of Violence* (Manchester, Manchester University Press, 1993).

[83] R Roy, 'Modern Economics and the Good Life: A Critique' (1992) 17 *Alternatives* 381–403.

[84] T Hobbes, *Leviathan* (New York, EP Dutton, 1950) 79–80.

[85] Roy, above n 86.

[86] Cited in Roy, above n 83, 372.

[87] R Eckersley, *Environmentalism and Political Theory: Toward an Ecocentric Approach* (New York, State University of New York, 1992) 18.

of human autonomy or self-determination. It also calls for a re-evaluation of the foundations of, and the conditions for, human autonomy or self-determination in Western political thought. Leiss[88] made a related point in his earlier writing when he wrote that everything depends not so much upon the establishment that limits to economic and population growth do exist but upon whether humans regard such limits as a bitter disappointment or as a welcome opportunity to turn from quantitative to qualitative improvement in the course of creating a conserver society. Democracies turn on the question of individual freedom, and never more so than when the sustainable development agenda gets personal, around the challenge of sustainable consumption. These are questions confronting advanced liberal democratic states and their citizens. The demands of sustainable consumption and the implicit critique of consumerism has begun to reopen questions about the State, and the relationship between the State and the individual.

At the level of the individual or self, the negotiation of freedom is also framed by the power and influence of powerful market actors and 'imagineers' with intimate access to the privatised desires of men, women and children. The influence operates largely in one direction, the direction of accumulation, with meaningful information about the consequences of consumer choices expelled to the margins. Barber[89] has argued that public liberty demands public institutions that permit citizens to address the public consequences of private market choices. He recalls that liberty has been understood as the capacity to make public choices (in Rousseau's terms to engage in 'general willing'), as a potential *faculty that must be learned* rather than a natural one that is exercised from birth:

> Rights are certainly moral claims but their effective exercise rests on competence and hence on learned skills of citizenship. That is why Tocqueville spoke of a necessary 'apprenticeship of liberty', which he called the most arduous of all apprenticeships.[90]

Prompted by the IPCC Fourth Assessment Report, Governments, including that of the UK, have begun to give serious attention to partially devolving responsibility for managing their emissions reduction targets to individual 'consumer-citizens'. Individual responses to initiatives such as Personal Carbon Trading will be framed by shifts in governance, which have seen risk and responsibility shifted onto individuals regarded as 'entrepreneurial selves'. This development is not a zero-sum game, and offers prospects for innovation, where individuals maintain or reclaim the capacity to act as 'citizens' alone and collectively where they choose to organise and use a PCT system to decommission allowances.

[88] Leiss, above n 4.

[89] B Barber, *Consumed: How Markets Corrupt Children, Infantilize Adults, and Swallow Citizens Whole* (London, WW Norton Company, 2007).

[90] Ibid 126–27. The reference to Alexis de Tocqueville is taken from his *Democracy in America* 1 (New York, Vintage Books, 1990).

The question of the relationship between consumerism, the role of the 'consumer-citizen', freedom and the legitimacy of the liberal state will emerge as one of the more critical questions around the State's response to climate change and other environmental challenges. Bauman[91] and Offer[92] have both offered insights into this problematic relationship, which is also tied up with the observations of Rose[93] and others on the transformation of governance of the liberal state under the sign of neoliberalism. The economics-driven neoliberal framework introduces a note of ambivalence to any claim that consumers might be afforded an opportunity to reconstruct an ethical dimension of their citizenship through this prospect of an enabling policy framework for individual carbon budgeting. Nevertheless, the prospect provokes further rethinking of the role of the State in its relations with the citizen and, more specifically, the co-dependent relation of State and consumer in an economy where the commanding social logic remains economic growth (facilitated by, inter alia, government-backed consumer debt facilities). The current financial crisis has already provoked a deep questioning of the Government's fiscal responsibilities to the citizen (and future generations).

Anticipating a new prominence for the carbon market and lifestyle change after the UNFCCC/Kyoto Protocol COP/MOP, in post-2012 implementation of ambitious carbon mitigation targets in OECD countries, including the UK, this chapter has drawn on Foucauldian scholarship to support an argument that PCT may be consistent with a new understanding of the role of the 'consumer citizen' in modern Britain, and that PCT may also come to be regarded as an advanced tool for triggering and supporting some of the elusive behaviour change required by the demands of sustainable consumption in the context of climate change. The chapter demonstrates that Personal Carbon Trading sits at the apex of a number of debates of interest to the environmental policy community, including climate change, sustainable consumption and individual freedom in an era of environmental constraints. These debates will tap into parallel discussions provoked by the global financial downturn, which has exposed some of the weaknesses of financial regulation and unsustainable levels of support for consumer credit, and exposed the extent to which the intimate choices of the consumer-citizen are already laden with risks largely mediated by globalised circuits of production and consumption.

[91] Z Bauman, above n 63.
[92] Offer, above n 12.
[93] Rose, above n 71.

12

State Responsibility for the Adverse Impacts of Climate Change on Individuals: Assessing the Potential for an Interdisciplinary Approach

MATTHEW HALL

INTRODUCTION

REFLECTING THE INTERDISCIPLINARY ambitions of this volume as a whole, this chapter draws together two apparently distinct fields: victimology (the study of victims and victimisation)[1] and international environmental law concerning climate change. The argument is made that, by combining aspects of both disciplines, progress can be made in both fields towards the recognition of individuals (in this case taken to mean natural persons) affected by the harmful effects of climate change and the development of mechanisms through which they might seek reparation from the *states* whose actions or inactions lead to such harm.[2] The value of such a dual perspective is demonstrated in the chapter first through an examination of the more traditional mechanisms by which states could be held responsible for the impacts of climate change on individuals, and secondly (and more specifically) through an application of human rights principles to both climate change and victims of crime more broadly. In conclusion, the chapter will set out a combined research agenda, drawing contributions from both international environmental law and victimological thought, in an effort to enhance each field both academically and in practice.

[1] Rather than entering into debates concerning victimology's status as an individual 'discipline' in its own right, the term is adopted here simply as a convenient and well-used descriptor for such investigations within criminology.

[2] Referred to as 'the victims of climate change' throughout this chapter. Of course, while this chapter is focused on individual (natural) persons as victims, the full ambit of such victimisation can extend much further to encompass corporate, community and other victims.

The principal contention of this piece is that closer collaboration between international legal scholars and criminologists interested in victimisation has the potential to markedly advance our understanding of a still somewhat under-researched issue: that of a state's responsibility for the adverse impacts of climate change on individuals. Such responsibility might exist both internally, where the actions or inactions of a state lead to harm for its own citizens, and also externally, where such harm occurs outside the state in question. This latter scenario derives from the fact that one of the foremost difficulties posed by climate change is that associated harms often occur in one state when the acts or omissions (lawful or otherwise) leading to climate change occur in a separate state. To pick one prominent example, the people of the Maldives are presently facing significant risk to their homes, economy and traditional ways of life as a result of sea level rises ostensibly brought about by climate change, despite the fact that the country itself is amongst the lowest contributor to the problem. Indeed, the unequal impact of climate change is well recognised by the international legal order, with the preamble to the 1992 UN Framework Convention on Climate Change (FCCC) acknowledging the particular vulnerability of 'low-lying and other small island countries, countries with low-lying coastal, arid and semi-arid areas or areas liable to floods, drought and desertification, and developing countries with fragile mountainous ecosystems'.[3] While this chapter concentrates the majority of its attention on the harm caused internally within a state, this wider question of state responsibility for external harm to individuals in another state poses yet another significant challenge for the international legal order. The added value of collaboration between disciplines, it is submitted, lies in the cross-fertilisation of structural ideas from the field of international environmental law (concerning the operation of international redress mechanisms, state liability, and human rights) and the more agency-driven ideas which inform much of victimology (focusing on the place of the *individual* within the quantum of harm brought about by climate change).

The argument could be made that the focus of this chapter on the individual *human* victims of climate change represents a limited, anthropocentric approach to a far wider problem.[4] The chapter should not however be read as downplaying the importance of other aspects of what White has termed 'ecological justice'.[5] The goal is rather to promote interdisciplinary dialogue, concerning what is still an underdeveloped area in both victimology and

[3] See also Art 4(8).

[4] The argument that international law is inappropriate for dealing with such individualistic suffering will be considered below.

[5] R White, *Crimes Against Nature: Environmental Criminology and Ecological Justice* (Cullompton, Willan Publishing, 2008) 18.

international environmental law, as part of the wider discussion on climate change and its impacts.

APPROACHING THE PROBLEM OF CLIMATE CHANGE

In recent years, climate change has become an accepted reality for most scholars and practitioners working in both the physical and social sciences.[6] The progress made in our understanding of the causes of climate change has raised a number of challenges for lawyers in particular, as questions are increasingly raised concerning the responsibility of individuals, corporations and states for environmental harms. Given the transboundary nature of the issue, international law in particular has been obliged to adapt itself rapidly to meet these new challenges, with the development of international environmental law.[7] There is however a fundamental difficulty in applying international law to the harms to individuals brought about by climate change, in that the international legal order is traditionally conceived as one devised by states, for states.[8] This approach leaves little room for the concerns of those *individuals* actually suffering harm within states, whether that harm comes in the form of physical or health-related issues or the broader loss of traditional ways of life, means of economic sustenance or territory as a result of climate change.

To traditional adherents of the positivist view of international (environmental) law, this exclusion of individuals is entirely proper and consistent with the understanding that only states (and perhaps a limited number of state-like entities) are the *subjects* of international law, whereas individuals are merely its *objects*.[9] The space is lacking in the present chapter to set out an extensive counterargument to this position.[10] It is nevertheless submitted that in light of the developments set out by McCordquodale[11] and others[12] concerning the recent application of individual rights at the

[6] N Oreskes, 'The Scientific Consensus on Climate Change: How Do We Know We're Not Wrong?' in J DiMento and P Doughman (eds), *Climate Change: What It Means for Us, Our Children, and Our Grandchildren* (Cambridge, MA, MIT Press, 2007) 68.

[7] See C Redgwell, 'International Environmental Law' in M Evans (ed), *International Law*, 3rd edn (Oxford, Oxford University Press, 2010) 687–721, 689–92.

[8] See P Alston, *Non-State Actors and Human Rights* (Oxford, Oxford University Press, 2005).

[9] L Oppenheim, *International Law* vol 1 (London, Longmans, 1905).

[10] See R McCordquodale, 'The Individual and the International Legal System' in M Evans (ed), *International Law*, 3rd edn (Oxford, Oxford University Press, 2010) 285–310.

[11] Ibid.

[12] See A Orakhelashvili, 'The Position of the Individual in International Law' (2007) 31 *California Western International Law Journal* 241–76 and C Ochoa, 'The Individual and Customary International Law Formation' (2007) 48 *Virginia Journal of International Law* 119–86.

international level—in areas which include economic law,[13] the law of self-determination[14] and human rights[15]—the exclusion of the individual from the ambit of international environmental law is looking more and more antiquated.[16] This is particularly true given the recent application of human rights to environmental issues, which will be discussed below. As such, Higgins[17] has suggested an alternative to the *subject* v *object* dichotomy in international law by focusing instead on the degree of 'participation' states and individuals have in international decision-making. Indeed, it will be discussed below how the concept of participation has become central to debates going on in both international environmental law and victimology. Even if one retains the traditional *subject v object* division the International Court of Justice has stated, in the context of ruling that the UN is a body with international personality, that the recognition of 'subjects' of international law must adapt to the present 'needs of the [international] community'[18] and is 'influenced by the requirements of international life'.[19] It is submitted that the growing international problems caused by climate change qualify in both respects as necessitating a serious rethink of *what* and *whom* we recognise in the international legal order, what rights they should expect and how the international legal order responds to those expectations.

While reference to individuals in international law was relatively scarce prior to the more recent incorporation of human rights within the field,[20] victimologists have long concerned themselves with the (usually criminal) harms perpetrated against individuals, arguably to the exclusion of a more structural and theoretically-informed debate on how such victims could achieve official recognition by national and international justice agencies.[21] Although much greater progress has been made along these lines is relation to domestic victims of 'traditional' criminal activities,[22] the victimological literature on the impacts of climate change is extremely brief and limited to a few key commentators. While the application of victimology to environmental harms as a whole was spearheaded in the

[13] A Lowenfield, *International Economic Law* (Oxford, Oxford University Press, 2008).

[14] A Buchanan, *Justice, Legitimacy and Self-Determination Moral Foundations for International Law* (New York, Oxford University Press, 2007).

[15] A D'Amato, 'Human Rights as Part of Customary International Law: A Plea for Change of Paradigms' (1996) 25 *Georgia Journal of International and Comparative Law* 71–92.

[16] It will be discussed below how international law has attempted to address this omission by reference to human rights.

[17] R Higgins, 'Conceptual Thinking About the Individual in International Law' (1978) 4 *British Journal of International Studies* 1–10.

[18] 'Reparation for Injuries Suffered in the Service of the United Nations (advisory opinion)' (1949) *ICJ Reports* 174, 178.

[19] Ibid.

[20] On which see below.

[21] S Walklate, *Imagining the Victim of Crime* (Maidenhead, Open University Press, 2007).

[22] M Hall, *Victims of Crime: Policy and Practice in Criminal Justice* (Cullompton, Willan Publishing, 2009).

mid-1990s by the work of Williams, his call for the development of what he then termed an 'environmental victimology'[23] has so far gone largely unanswered. Nigel South has also reflected upon the unequal impact of climate change on various groups of (usually poor) victims, but has not expanded this discussion into a full-length piece.[24]

This state of affairs is somewhat surprising given the origins of the so-called 'victims' movement', driven as it was by a host of 'strange bed-fellows'[25] concerned with different aspects of victimisation in its broader sense (beyond criminal victimisation). These ranged from feminists and mental health practitioners, to survivors of war and atrocities such as the Nazi concentration camps and natural disasters.[26] The failure of modern victimology to engage with the issue of environmental harm is partly explained by the field's recent preoccupation with *criminal* victimisation and the position of such victims within the justice system.[27] This being the case, victimologists have been slow to consider the victims of climate change not only because scientific knowledge on the issue is only now reaching a consensus point, but because many of the activities which foster climate change constitute entirely legal activities at both the national and international levels.[28]

Of course, affording rights (or indeed responsibilities) to victims under international environmental law is not necessarily the same as allowing them to bring claims against polluting states in the international legal order for the harmful effects of climate change. Indeed, it can be argued that such victims might be better off targeting polluting corporations under private international law rather than taking on a state, which is conceptually and historically more problematic. On this issue the victimological literature consistently suggests that victims, having suffered harm, greatly value receiving some form of direct redress or apology from the party responsible for that harm, over and above the official recognition implied by a (domestic) criminal conviction or other official form of holding to account, and even when this does not 'compensate' for the harm suffered

[23] C Williams, 'An Environmental Victimology' (1996) 23 *Social Science* 16–40 reprinted in R White (ed), *Environmental Crime: A Reader* (Cullompton, Willan Publishing, 2009) 200–22, 202.

[24] N South, 'The Ecocidal Tendencies of Late Modernity: Trans-National Crime, Social Exclusion, Victims and Rights' in R White (ed), *Global Environmental Harm: Criminological Perspectives* (Cullompton, Willan Publishing, 2010) 236–40.

[25] M Young, 'Ideological Trends Within the Victims' Movement: An International Perspective' in R Davis, A Lurigio and W Skogan (eds), *Victims of Crime*, 4th edn (Thousand Oaks, Sage Publications, 1997) 115–26, 120.

[26] Ibid.

[27] M Maguire, 'The Needs and Rights of Victims of Crime' in M Tonry (ed), *Crime and Justice: A Review of Research* (Chicago, Chicago University Press, 1991) 363–433.

[28] White, n 5 above.

in strict economic terms.[29] In other words, receiving reparation *from offenders* has important symbolic and cathartic benefits for victims of crime. Nevertheless, the important point is that polluting activities which lead to climate change are often brought about via activities sanctioned *by the state*, either proactively (through granting permits or licenses to carry out hazardous activities) or by omission in failing to regulate such activities tightly. Furthermore, private companies may not be in a position to provide full or prompt compensation to such victims, a fact acknowledged by the International Law Commission.[30] As such, it is the *state* which may ultimately be responsible for environmental harms brought about by climate change, particularly when the activities carried out by polluting corporations are entirely legal.

In domestic criminal matters, the most common mechanism through which a state is obliged to support victims who have been harmed are the various state compensation schemes set up in most developed nations. In the majority of these schemes victims of certain classifications of crime (usually violent crime) are able to apply to the state to compensate them for physical and psychological injuries.[31] Significantly, these schemes usually work on a no-fault basis.[32] That is, they are based on welfare ideals rather than any admission on the part of the state that it is somehow 'responsible' for the harm caused to individuals by 'failing' to protect them or to prevent crime.[33]

In Europe, this principle has been applied at the transboundary level in EU Directive 2004/80, relating to compensation to crime victims, which requires Member States to co-operate in providing compensation to victims of crime victimised on their soil but resident in another Member State, such that the victim may apply directly to their home state compensation mechanism but receive payment from the state where the crime occurred. Notably, however, the Directive makes it clear that the *harm* to the victim must have occurred in the latter state, meaning it cannot apply to harm occurring in a home state caused by actions in another state.[34] In any event, it seems unlikely that states will agree to start compensating victims of environmental harm

[29] J Shapland, 'Bringing Victims in From the Cold: Victims' Role in Criminal Justice' in J Jackson and K Quinn (eds), *Criminal Justice Reform: Looking to the Future* (Belfast, Queen's University Belfast, 1990) 1–10.

[30] International Law Commission, 2006 Draft principles on the allocation of loss in the case of transboundary harm arising out of hazardous activities, with commentaries (Report on the work of the fifty-eighth session (1 May to 9 June and 3 July to 11 August 2006) General Assembly Official Records Sixty-first Session Supplement No. 10 (A/61/10)). Principle 4(5) and Principle 7(2). The draft principles are discussed in more detail below.

[31] D Miers, *Compensation for Criminal Injuries* (London, Butterworths, 2001).

[32] As do the ILA's draft principles, n 30 above, Principle 4(2).

[33] Albeit Miers (n 31 above) argues that the true rational behind state compensation schemes is often difficult to pinpoint.

[34] Art 2.

in their own territory, let alone in other countries, on the same welfare basis as they pay domestic victims of crime.

As noted above, the International Law Commission (ILC) has responded to the issue of compensation with draft principles on the allocation of loss in the case of transboundary harm arising out of hazardous activities. In particular the principles call for states to 'take all necessary measures to ensure that prompt and adequate compensation is available for victims of transboundary damage caused by hazardous activities located within its territory or otherwise under its jurisdiction or control'.[35] While it is generally envisioned that such compensation should come from the 'operator' in command or control of the relevant activity at the time of the incident leading to transboundary damage, as previously noted the principles also maintain that states themselves must supplement funds in the event that operators are unable to provide full compensation to victims.[36] Also of note is the provision in Principle 6(2) for victims in states beyond that where hazardous activities are located:

> Victims of transboundary damage should have access to remedies in the State of origin that are no less prompt, adequate and effective than those available to victims that suffer damage, from the same incident, within the territory of that State.

The above notwithstanding, the ILC principles are presently only in draft form and are centred on receiving compensation from 'operators' below the state level. There is as yet no established international mechanism for victims of environmental harm to receive compensation from a polluting state. This effectively leaves such victims with the significant task of demonstrating some form of legal responsibility on behalf of such a state. As such, it seems that victimologists must turn to a complex array of international and transboundary (private international law) rules if they are to ascribe the victims of environmental harm with rights in any justice process that ensues from polluting activity.

Adding Value: an Exchange of Agency and Structure?

As noted above, international environmental law as presently conceived generally lacks deference to individual harm, or the agency of individuals affected by that harm. Agency here is given the conventional sociological meaning as a synonym for undetermined human action.[37] In traditional sociology agency is contrasted with structure, which broadly reflects

[35] Principle 4(1).

[36] Principles 4(5) and 7(2).

[37] N Abercrombie, S Hill and B Turner, *The Penguin Dictionary of Sociology* 5th edn (London, Penguin, 2006).

recurring social processes, institutions and rules which limit the action of individuals and to some extent determines them.[38] Structure therefore generally includes the legal systems put in place by society.[39] This duality informs much of traditional criminology, being the essential bedrock of debates between the rational actor and deterministic views of criminality. Of greater significance to the present discussion however are the more recent debates concerning the interrelationship, and possible interdependence, of these concepts. One of the most influential writers to offer such arguments is Antony Giddens, who champions the concept of a 'duality of structure'[40] whereby the structural properties of social systems are produced and reproduced by social practices (ie by human agency). Bourdieu too proposes a model of closer integration between structure and agency, essentially arguing that individuals become disposed to act in a manner which promotes certain codes of behaviour and their associated social structures.[41] Hence, structure becomes embedded within agency over time. More recently Mouzelis has commented and expanded upon both Gidden's and Bourdier's perspectives to incorporate the changing of social structures through agency (not just their production and reproduction as per Giddens) and also for a 'rational, calculative and reflective social actor'[42] endowed with greater freedom of action than Bourdieu recognises.

The full detail of these debates is beyond the scope of the present discussion. The important point is that the previous dualistic conceptions of structure and agency, whereby one essentially champions the cause of one or the other as *the* principal driver of social phenomena (including law and legal systems), has been replaced by the understanding that both, variously configured, are necessary to understand society. This returns us to the critique made earlier of the traditional international legal order regarding climate change and its denial of individual (or indeed group) human agency. At an operational level we have seen how many international lawyers defend this position by reference to the supremacy of the state in the international legal order. This structural straightjacket has lead to a dearth of commentary amongst international environmental lawyers on the issue of climate change *which also encompasses human agency*. If Giddens, Bourdier and Mouzelis are correct, a full understanding of such a problem can never be reached with such limitations. Here then lies the added value of victimological discussion in this debate, as victimology is very much concerned with individuals. Victimology is also of course concerned with social

[38] Ibid.

[39] T Burns and H Flam, *The Shaping of Social Organization: Social Rule System Theory With Applications* (London, Sage Publications, 1987).

[40] A Giddens, *The Constitution of Society* (Cambridge, Polity Press, 1984) 85.

[41] P Bourdieu, *The Logic of Practice* (Cambridge, Polity Press, 1990).

[42] N Mouzelis, *Modern and Postmodern Social Theorising* (Cambridge, Cambridge University Press, 2008).

structures and the place of victims within them, particularly criminal justice systems.[43] Nevertheless, it will be shown below that the general lack of enforcement structures for victims' rights at the national and international level, coupled with the underdevelopment of theory in relation to victim rights,[44] means international law has also much to offer victimology in the development of a more structurally-nuanced position.

In sum, it is submitted that international environmental lawyers can benefit from the more individualistic, agency-based, discussion provided by the victimological literature (incorporating the *people* who, after all, constitute states), whereas victimologists can draw from the structural frameworks developed by international lawyers as a means of furthering official recognition and redress for victims of harms brought about by climate change (and perhaps other harms). Such an approach not only benefits from the best of both academic traditions, but also conforms to modern theoretical notions in social science to the effect that structure and agency are often inherently linked together, such that they cannot be considered in isolation.[45] The remainder of this chapter will be devoted to providing examples of such an exchange of ideas in practice, and how such an exchange may help develop international law in relation to climate change. This will begin with the question of state responsibility for climate change and its associated harms, before moving on to a discussion of human rights and their application to the problem.

STATE RESPONSIBILITY FOR THE IMPACTS OF CLIMATE CHANGE

A number of arguments have been put forward in favour of states taking legal responsibility for the impacts of climate change, albeit few have won universal acceptance as principles of international law. This latter point is significant because, in the general absence of legally binding multilateral rules determining state responsibility for climate change,[46] such responsibility can only be established as a binding principle of *customary* international law if it can be shown to constitute a common practice by states, with particular regard to those states most affected by that practice.[47]

[43] J Shapland, J Willmore and P Duff, *Victims in the Criminal Justice System* (Aldershot, Gower, 1985).

[44] With some notable exceptions, including Walklate, above n 21, and P Rock, 'Theoretical Perspectives on Victimisation' in S Walklate (ed), *Handbook of Victims and Victimology* (Cullompton, Willan Publishing, 2007) 37–61.

[45] See C Shilling 'Reconceptualising Structure and Agency in the Sociology of Education: Structuration Theory and Schooling' (1992) 13 *British Journal of Sociology of Education* 69–87.

[46] Redgwell, above n 7.

[47] Statute of the International Court of Justice, Article 38 1(b). See also H Thirlway, 'The Sources of International Law' in M Evans (ed), *International Law* 3rd edn (Oxford, Oxford University Press, 2010) 95–121.

This section will demonstrate how the already complex problem of establishing responsibility on the part of states for climate change itself becomes even more challenging when an attempt is made to attribute that responsibility to harm suffered by individuals, affording such individuals some form of redress. Again the contention is that this represents a significant and ultimately restrictive disregard for the role of agency in the formation of legal systems.

In terms of substantive instruments, the key response of the international community to the problem of climate change is the 1992 Framework Convention on Climate Change (FCCC), which embodies obligations for states to take action to mitigate its adverse affects. The 1997 Kyoto Protocol to the Framework Convention sets legally enforceable emissions targets for countries. Some commentators have also suggested that the Kyoto Protocol might embody *obligations erga omnes*, or obligations which are owed to the international community as a whole by all states.[48] In theory any state can seek redress in the event that such obligations are breached, although there is little evidence that such an obligation relating to climate change has been accepted, as a principle or as a custom, into the corpus of international law.

The above paragraph notwithstanding, Kyoto is not universally applicable given that, significantly, the United States has stated that it will never ratify the Protocol. Therefore the US, as the world's biggest polluter per capita, is not bound by any emissions reduction targets. Arguably the US still has an obligation to limit its emissions of carbon and greenhouse gases because it remains a *party* to the FCCC and, under Article 18 of the Vienna Convention on the Law of Treaties, a party to a treaty is under an obligation not to do anything that would defeat the object and purpose of a treaty. Perhaps the more significant impact of the US failure to ratify Kyoto is the implications this has for customary international law. That is to say, it becomes somewhat unconvincing to argue that the obligations contained in the FCCC and in Kyoto constitute the required generality of practice when the world's biggest polluter has rejected them. Moreover, the climate change omission targets within the Kyoto protocol are based on a principle of shared but different responsibilities between countries. As such, while no party to the Protocol is entirely excused responsibility for climate change, developing countries—more than half of the world's nations—do not face the same level of omissions targets, including major emitters like China and India. As such, although the need to take action to militate against climate change is morally recognised by most states, climate change obligations are not in themselves customary international law.

[48] See R Higgins, 'Conceptual Thinking About the Individual in International Law' (1978) 4 *British Journal of International Studies* 1–10.

Significantly for the present discussion, neither of the above instruments include provisions expressly aimed at the individuals harmed by climate change. The FCCC acknowledges in its preamble that climate change is 'a common concern of humankind' and in Article 1(1) adopts a definition of the 'adverse effects of climate change' which includes 'significant deleterious effects on human health and welfare'. As such Article 4(1)(f) of the Framework Convention requires parties to minimise the adverse impacts of climate change on public health. Article 3(1) also speaks of protecting the climate system 'for the benefit of present and future generations of humankind'. Notably all of these references are collective, rather than specific to particular groups of citizens, or individuals, affected by climate change. This very much reflects the state-centred paradigm in international environmental law, as in international law in general.[49]

Other attempts to identify responsibilities of states concerning climate change are based around broader principles which have been argued by some to have become implicit in customary international law. For example, Principle 21 of the 1972 UN Declaration on the Human Environment (known as the Stockholm Declaration) imposes upon states the obligation to ensure that activities within their jurisdiction or control do not cause transboundary harm. This standard was also contained in Principle 2 of the 1992 Rio Declaration on Environment and Development. These are non-binding instruments, although there is a body of supporting evidence to suggest that the no-harm principle at least has entered into the corpus of customary international law, tracing back to the seminal judgment made against Canada by the Mixed Arbitral Tribunal[50] in the *Trail Smelter Arbitration* concerning cross-border air and water pollution.[51] In this case the arbitrator of the Tribunal postured that 'A State owes at all times the duty to protect other states against injurious acts by individuals within its jurisdiction'.[52] In addition, the International Court of Justice in the *Corfu Channel Case* has pronounced that a state is under an obligation not to 'knowingly allow its territory to be used for acts contrary to the rights of other states'.[53] As noted by Redgwell, however, the no-harm principle is limited in that it constitutes only a negative obligation on states not to allow their territories to be used for activities which cause harm to the

[49] Although it should be noted that the International Law Commission's draft principles, n 30 above, Principle 2(f).

[50] Constituted by the 1936 Convention of Ottawa between the Canada and the US.

[51] Trail Smelter Arbitration Tribunal (*US v Canada*) 33 AJIL 182 (1939) and 35 AJIL 684 (1941).

[52] Trail Smelter Arbitration Tribunal (*US v Canada*) 33 AJIL 182 (1939) and 35 AJIL 684 (1941), Article XII, citing C *Eagleton Responsibility of States in International Law* (New York, New York University Press, 1928) 80.

[53] *Corfu Channel Case* (1949) ICJ Reports 4, 22.

environments of other states, or the global commons.[54] This falls far short of a positive obligation to protect the environment.[55] There is also no real or substantial attempt in any of these pronouncements to recognise the harm caused to *individuals* within states. Again, 'harm' here seems to mean harm against another state in general.

Redgwell has labelled other general principles which have been applied to climate change as 'even more controversial' in terms of their status as customary law.[56] One prominent example is the so-called 'precautionary principle', this being the belief that society should seek to avoid environmental damage through forward planning and blocking the flow of potentially damaging activities despite an absence of full scientific certainty on the matter. The principle is found under the Rio Declaration[57] and has the advantage that it is able to operate despite scientific uncertainties as to the causes of climate change. That said, being adopted more in Europe than the US, again there seems to be a lack of universal application for this to be recognised as customary international law.[58] In addition, from an evidential perspective, if there is no scientifically proven relationship between specific acts/omissions of states and the harms caused to victims this would make it very difficult to establish specific responsibility or compensation claims on this basis.

Other general principles which have been applied to climate change include that of *sic utere tuo, ut alienum non laedas* or 'use your own property in such a way that you do not injure others', which was incorporated into Principle 21 of the Stockholm Declaration in 1972, and the principle of sustainable development, which is stated in Principle 3 of the Rio Declaration as:

> The right to development must be fulfilled so as to equitably meet developmental and environmental needs of present and future generations.

While this pronouncement does incorporate the human element, as with the FCCC it does so only in the forward-reaching, collective sense of humankind and intergenerational justice, rather than acknowledging human agency in the here and now. In this respect the so-called 'polluter pays' economic principle seems to have greater potential for attributing some degree of responsibility for the harms caused to individuals by pollution to the state. The thrust of this principle is that the wrongdoer (polluter)

[54] Redgwell, above n 7. The author cites the 1982 United Nations Convention on the Law of the Sea and the 1959 Antarctic Treaty as arguably containing rare examples of positive obligations to protect the environment.

[55] Although there is judicial precedent to support the counterargument. Notably by Judge Weeramantry in his dissenting opinion in *Legality of the Threat or Use of Nuclear Weapons, Advisory Opinion* 1 (1996) ICJ Reports 226 at Part 3 sec 4:1(b).

[56] n 7 above at 695.

[57] Principle 15.

[58] D Wybe, *The Precautionary Principle* (The Hague, TMC Asser Institute, 1997).

is under the obligation to make good the damage caused. The principle has already become widely accepted as a means of paying for the cost of pollution and domestic control,[59] and is of particular interest in the context of the present discussion because it tends to encompass the public interest, as in Principle 16 of the Rio Declaration:

> National Authorities should endeavour to promote the internalization of environmental costs and the use of economic instruments, taking into account the approach that the polluter should in principle bear the cost of pollution with due regard to the public interest and without distorting international trade and development.

The difficulty though is that the polluter pays principle has so far not been applied to *nation states* as responsible parties for pollution.[60] This is in contrast to the no harm principle which, while aimed at member states, does not concern itself with the harm caused *to individuals*, but rather to other states. In both cases then the established principles fail to join the nation-state as a responsible party with the individual victims who are harmed. Again, there is a lack of unified consideration of structure and agency.

To summarise the above section, some 70 years after the *Trail Smelter Arbitration* provided such high hopes for those seeking to assert the responsibility of states for environmental harm, including climate change, the difficulties of establishing such responsibility are still great. While progress has been made both in terms of substantive treaty-based international law and in terms of general principles in the international legal community, it is generally still difficult to argue that most such rules have been accepted as universal principles of international law. The no-harm principle may be an exception, although the 'harm' conceived here clearly suggests harm caused to states rather than to individuals within states. Our understanding of 'harm' is a key component of the following section, which draws on victimological literature in an attempt to widen this restrictive approach.

Of course, in legal terms, establishing state *responsibility* for climate change is only a start. Once an obligation has been established in principle there is the need to demonstrate conclusively that a state has breached that obligation and then causation would need to be established between the actions of that state and the harm caused to the individual in another state. There is insufficient space in this chapter to examine these significant legal hurdles in more detail but it can be confidently predicted that, while it has been noted that states have taken to compensating home victims of violent crimes without necessarily accepting fault, it is extremely unlikely that they will do the same for those affected by climate change in their own

[59] J Tobey and H Smets, 'The Polluter-Pays Principle in the Context of Agriculture and the Environment' (1996) 19 *World Economy* 63–87.

[60] Ibid.

or other countries,[61] not least because this would have very different cost implications.

THE MISSING VICTIM?

The above discussion suggests that, even if states are, through some mechanism, principle, or custom of international law, held to be responsible for climate change (itself a difficult proposition) it seems even less likely at present that they can be held responsible for the impacts *of* climate change as they pertain to individual citizens, or even groups of citizens defined in more specific terms than 'humankind' or 'subsequent generations'. In sum, the traditional positivist view of international law as state-oriented remains as dominant in matters of climate change as it does in most other areas. The key question for this chapter then becomes whether victimology has anything to contribute towards moving these debates forward.

In one of the few dedicated works in victimology concerning environmental harm, Williams[62] has argued that the definition of victims of abuse of power found under the 1985 UN Declaration of Basic Principle of Justice for Victims of Crime and Abuse of Power (hereafter the 1985 UN Declaration) can serve as a useful starting point for ascribing rights to environmental victims. The definition covers:

> [P]ersons who, individually or collectively, have suffered harm, including physical or mental injury, emotional suffering, economic loss or substantial impairment of their fundamental rights, through acts or omissions that do not yet constitute violations of national criminal laws but of internationally recognized norms relating to human rights.[63]

The 1985 UN Declaration is the most wide-ranging and influential document on the issue of victims' rights at the international level. The Declaration speaks of victims being afforded access to justice, fair treatment and of compassion and respect for their dignity. While an essentially non-binding, soft-law instrument, the Declaration has in turn influenced most of the domestic provisions relating to the place of victims in criminal justice (which usually take the form of non-binding Codes of Practice) in the world today.[64] Nevertheless, in line with the above criticisms concerning the general focus of the victim movement on human agency as opposed to structure, the Declaration is very much lacking in mechanisms for the enforcement of the principles it champions (at the national and international levels), and is in fact a self-confessed aspirational document.

[61] Despite the provisions of the ILC draft principles, n 30 above.
[62] Williams, above n 23, 202.
[63] Art 18.
[64] Hall, above n 22.

The inclusion of victims of abuse of power within the 1985 Declaration was intended to encompass victimisation by the *state*,[65] making this definition particularly relevant from the perspective of victims of climate change in the context of the present discussion. Kauzlarich *et al*[66] have drawn on this provision in the development of a framework 'victimology of the state', which includes 'International-International Governmental Crime', or crime which occurs outside a state's geographic jurisdiction against international law or human rights. Such an understanding would encompass victimisation not only by one's *home* state, but by other states as well. It is particularly significant that the definition given for victims of abuse of power includes actions and omissions which do not constitute crimes because, as noted above, the activities leading to climate change are invariably not illegal in nature. More specifically, the inclusion of *omissions* under this definition of victims of abuse of power is particularly significant for present purposes; especially when combined with the no-harm principle discussed above. Thus, the case could be made that a state is responsible under the 1985 Declaration for failure to regulate activities (that is, failure to show 'due diligence'[67]) within its borders which lead to transboundary environmental harm, thus constituting an abuse of power. While it may seem fanciful to read victims of climate change into a document conceived long before the term was in popular use, or indeed the problem was widely accepted and understood, in domestic law at least there is precedent for such an adaptation in the form of the US Crime Victims' Rights Act of 2004 (CVRA).[68] Following the explosion of a BP Oil Refinery in Texas in 2005, the US Fifth Circuit court ruled that the government had violated victims' rights under the CVRA by failing to consult with those locals affected by the explosion (mostly in the form of personal injury and property damage) in the agreement of a plea bargain with BP.[69] This was despite the fact that the number of victims stretched into the hundreds and the CVRA neither includes nor, on a standard reading, conceives harm caused by environmental damage. This domestic application of victims' rights to environmental harm is admittedly several steps away from the acceptance of individual victims of climate change into international law, but it does demonstrate how instruments traditionally presumed to extend only to victims of 'traditional' domestic offences (and in this case one heavily based on the 1985 Declaration) can be

[65] C Reese, 'The Implementation of the UN Declaration of Basic Principles of Justice for Victims of Crime and Abuse of Power in France' (2000) 4 *The Victimologist* 1–2.

[66] D Kauzlarich, R Matthews and W Miller, 'Towards a Victimology of the State' (2001) 10 *Critical Criminology* 173–94.

[67] R Barnidge, 'The Due Diligence Principle Under International Law' (2006) 8 *International Community Law Review* 81–121.

[68] 18 USC § 3771.

[69] J Starr, B Flack and A Foley, 'New Intersection: Environmental Crimes and Victims' Rights' (2008) 23 *Natural Resources and Environment* 41–52.

applied to environmental harm. The critique here is not that international law has considered and rejected this argument in relation to climate change specifically, but rather that, through its championing of legal rules and process over substantive *outcome*, such arguments are rarely even considered.

The other crucial point concerning the 1985 Declaration's definition of victims of abuse of power is its centring on the concept of 'harm'. It is clear from the definition as a whole that harm in this context is defined widely and applies to the individual 'including physical or mental injury, emotional suffering, economic loss or substantial impairment of their fundamental rights'.[70] In other words the Article is concerned with the infliction of harm to individuals and groups, broadly defined, rather than 'violations of national criminal laws' per se. Again this reflects the championing by the victims' movement of individual agency over legal structure. From the perspective of climate change the significant implication of this is that abuse of power can include the grey legal areas alluded to earlier in which environmental destructive activities leading to climate change often fall. It is notable in this regard that the ILC's draft principles, discussed earlier, similarly draw on the concept of harm (or 'damage'), defining victims as 'any natural or legal person or State that suffers damage',[71] 'damage' here including loss of life, personal injury and financial costs attributed to environmental harm.[72]

On a theoretical level, this focus on harm is itself an important contribution from the criminological literature, where in recent years criminologists generally and victimologists specifically have been criticised for focusing on officially prescribed definitions of crime and victimhood, giving impetus to the development of so-called 'radical' schools in both criminology[73] and victimology.[74] It is this development of so-called 'radical victimology' which invites consideration of the victims of climate change and other environmental harms within this field, because the radical agenda encompasses 'real, complex, contradictory and often politically inconvenient victims'.[75] To this we might also add *economically* inconvenient victims, given that, as White has demonstrated, there are often strong economic incentives for individuals, businesses and governments to passively allow or actively pursue activities which lead to climate change victimisation.[76]

[70] Art 1.

[71] n 30 above, Principle 2(f).

[72] n 30 above, Principle 2(a).

[73] Kauzlarich *et al*, above n 66.

[74] J Jordan, 'Lest We Forget: Recognising and Validating Victims' Needs', Paper Presented at the *Addressing the Underlying Causes* of *Offending; What Is* the *Evidence*? Conference; University of Wellington (27 February 2009).

[75] T Kearon and B Godey, 'Setting the Scene: A Question of History' in S Walklate (ed) *Handbook of Victims and Victimology* (Cullompton, Willan Publishing, 2007) 17–36, 31.

[76] White, above n 5.

Of course, in relation to climate change this radical critique applies not just to the shortcomings of international environmental law, but to victimology itself, for while there has been some progress in utilising this radical framework to address the victims of white-collar crime generally[77] there has been very little discussion of its implications for the study of victims of climate change. More significantly for the present discussion, there is a reciprocal benefit to victimology from international law because it is the structures and principles of international law which could ultimately promote the Declaration and other such documents to the level of formal and binding legal rules and/or principles. Unlike most victimologists, international lawyers are well versed in arguments concerning the enforceability of their subject matter, which is often much more contestable than in most other areas of legal scholarship (where the basic legal force of an instrument is often a given). Thirlway describes how international lawyers deal much more regularly with so-called 'secondary rules' which determine the sources and authority of legal and quasi-legal rules.[78] International law can thus furnish victimology with a ready-made vocabulary in which to conceptualise such measures as well as presenting another dimension of the transnational and international policy networks connecting such reforms. For example, some international lawyers have championed the concept of 'soft law' which has the capacity to form the material basis of customary law[79] and can be converted into 'hard law' though its enshrinement in treaties.[80] The concept of soft law is greatly contested in the relevant literature.[81] In practice, whether or not one chooses to call the 1985 UN Declaration, or indeed the ILC draft principles, soft law has little bearing on arguments concerning its potential either to become hard law or constitute an important step in the process towards the establishment of internationally binding legal principles relating to victims of climate change. Indeed, d'Aspremont has reflected on the argument that, given the operation of policy networks, legally binding instruments are not always needed to achieve significant ends:

Many recent developments, like networks among governmental officials or transnational law, have shown that non-legal instruments may prove more adapted to the speed and complexity of modern international

[77] B Williams, *Working With Victims of Crime: Policies, Politics and Practice* (London, Jessica Kingsley Publishers, 1999).

[78] Thirlway, above n 47.

[79] Cheng has gone so far as to suggest that an appropriately worded UN resolution can create *instant* customary law: B Cheng, 'United Nations Resolutions on Outer Space: "Instant" Customary Law?' (1965) 5 *Indian Journal of International Law* 23–48.

[80] A significant case in point being the UN Universal Declaration of Human Rights of 1948, enshrined in 1966 in the International Covenant on Civil and Political Rights.

[81] A Boyle, 'Soft Law in International Law-Making' in M Evans (ed), *International Law*, 2nd edn (Oxford, Oxford University Press, 2006) 141–58.

relations and are more and more resorted to in practice. Non-legal instruments can be at least as integrative for a community as legal ones.[82]

Given the generally non-binding nature of most national and international provisions and principles relating to both climate change and victims as a whole, such insight is important for both fields of study. What victimology also contributes to these debates is a wealth of experience in dealing almost exclusively with such non-binding instruments in attempting to attribute rights or so-called 'legitimate expectations'[83] to victims. Such 'rights' are almost exclusively found in non-binding declarations and codes of practice issued by national governments and international organisations.[84] The important observation is that such instruments have brought about significant changes in many jurisdictions concerning the place of victims in criminal justice, notwithstanding their non-binding or persuasive character.[85] This again implies that 'hard law' is not the only mechanism for enhancing the position of victims of climate change, politically or legally.

HUMAN RIGHTS: A COMMON SOLUTION?

In light of the above difficulties in applying the traditional international legal order to individuals harmed by the impacts of climate change, some international environmental scholars have turned to the question of human rights as a means of attributing responsibility to states. Indeed Birnie and colleagues have recently described this development as 'perhaps the most significant shift in the focus of international law'.[86] As noted above, victimologists have increasingly moved in this direction as well in the face of similar legal hurdles both domestically and at the international level. Victims' rights is itself a complex and controversial proposition, victims of crime generally lacking the usual taken-for-granted rights enjoyed by defendants under most conceptions of natural (criminal) justice. Hence, while Ashworth refers to the 'normal' rights' of defendants in criminal law[87] both

[82] J d'Aspremont, 'Softness in International Law: A Self-Serving Quest for New Legal Materials' (2008) 19 *European Journal of International Law* 1057–93, 1062.

[83] JUSTICE Committee, *Victims in Criminal Justice, Report of the JUSTICE Committee on the Role of Victims in Criminal Justice* (London, JUSTICE, 1998) 36.

[84] With the possible expectations of the 2001 EU Council Framework Decision on the Standing of Victims in Criminal Proceedings (discussed below) and the US Crime Victims' Rights Act of 2004 (discussed above).

[85] M Hall, *Victims and Policy Making: A Comparative Approach* (Cullompton, Willan Publishing, 2010).

[86] P Birnie, A Boyle and C Redgwell, *International Law and the Environment* (Oxford, Oxford University Press, 2008) 269.

[87] A Ashworth, 'Victims' Rights, Defendants' Rights and Criminal Procedure' in A Crawford and J Goodey (eds), *Integrating a Victim Perspective Within Criminal Justice: International Debates* (Aldershot, Ashgate Dartmouth, 2000) 185–204.

Ashworth and Edwards have objected to any 'common sense' grounding of victims' rights.[88]

For international environmental lawyers, the application of human rights law in this area similarly remains distinctly underdeveloped; reflecting the continued state-oriented nature of the debate reflected above. A further level of complexity lies in debating the degree to which such rights should be purely anthropocentric, and the extent to which the rights of the individual should be weighed against any damage to the wider environment, ecosystem, etc.[89] As noted above, the focus in this chapter on individual (human) victims should in no way detract from the importance of these wider questions.

Two broad approaches have been taken to ascribing the victims of climate change (and other environmentally destructive activities) with relevant rights: theorising a new right to a clean and unpolluted environment (the less anthropocentric path) and arguing that such an environment, while not a distinct right in itself, is necessary in order to guarantee a more recognised human right such as, for example, the right to life or the right to privacy.[90] A similar distinction can be drawn in the victimological literature, and associated policy movements, between the development of specific victim-focused schemes—such as the rollout of victim impact statements in many jurisdictions,[91] the introduction of information services for victims of crime[92] and the general courteous treatment of such victims mandated by relevant (often non-binding) guidelines—on the one hand, and broader legal arguments concerning the application of existing human rights to victims as well as to defendants on the other.[93]

While considerable progress has been made in both fields in recent years towards the assimilation of human rights, there have also been significant challenges to what in both cases is a very novel application of the rights concept. Indeed, human rights have generally not been conceived as extraterritorial in nature.[94] As such, no right to a healthy or clean environment has yet been widely recognised by the international legal order. At a more local level one can find such rights articulated in the American Convention on Human Rights Protocol of San Salvador[95] and, more specifically, in the 1981 African Charter on Human and People's Rights as interpreted

[88] I Edwards, 'An Ambiguous Participant: The Crime Victim and Criminal Justice Decision-Making' (2004) 44 *British Journal of Criminology* 967–82.

[89] White, above n 5, at 15.

[90] The latter option is sometimes referred to as the 'greening' of human rights.

[91] E Erez, 'Who's Afraid of the Big Bad Victim? Victim Impact Statements as Victim Empowerment and Enforcement of Justice' (1999) July *Criminal Law Review* 545–56.

[92] B Williams, 'The Victims' Charter: Citizens as Consumers of Criminal Justice Services' (1999) 38 *Howard Journal of Criminal Justice* 384–96.

[93] A Sanders and R Young, *Criminal Justice* (London, Butterworths, 2000).

[94] D'Amato, above n 15.

[95] Art 11.

in the *Ogniland* decision of the African Commission on Human Rights and Peoples Rights. The *Ogniland* decision is particularly significant given the finding of the Commission that African states were obliged under the Charter to 'prevent pollution and ecological degradation, to promote conservation, and to secure ecologically sustainable development and use of natural resources'.[96] While only a local decision, conferring at best what can be seen as a collective right rather than an individual right, Birnie and colleagues have called this 'a remarkable decision which goes further than any previous human-rights case in the substantive environmental obligations it places on states.'[97]

More generally, however, while the right to a clean environment was suggested by the 1972 UN Conference on the Human Environment, and is reflected in Principle 1 of the Stockholm Declaration, the Rio Declaration of 1992 neglected to develop this further. Possibly the most progressive source of legally binding environmental obligations in the corpus of international law is found in the 1998 UNECE Convention on Access to Information, Public Participation in Decision-Making and Access to Justice in Environmental Matters (the 'Aarhus' convention), although this only provides victims of environmental harm with participation and procedural rights (to be discussed below), along with rights to information, as opposed to a blanket right to a clean environment per se. A right to a healthy and decent environment has also been suggested by the UN Sub-Commission on the prevention of Discrimination and Protection on the grounds that it would enhance the standing of environmental concerns when balancing conflicting rights and objectives,[98] although this recommendation was not taken forward.

Rights for victims of crime in general have received more explicit attention at the national and international level than the victims of climate change specifically. Nevertheless, a brief examination of the content of such rights can offer insight into issues the purely structural approach often taken by international environmental lawyers has often overlooked. The 1985 UN Declaration speaks of victims (of crime and abuses of power) being afforded access to justice, fair treatment and of compassion and respect for the dignity of victims.[99] The instrument also makes reference to victims being informed about their role and about the scope, timing, and progress of proceedings, as well as the disposal of their case.[100]

[96] *The Social and Economic Rights Action Center for Economic and Social Rights v Nigeria*, ACHPR, No 155/96 paras 52–53.

[97] Birnie *et al*, above n 86 at 273.

[98] See Sierra Club, 'Legal Defence Fund 'Human Rights and the Environment: the Legal Basis for Human Rights to the Environment' Report to the UN Sub-Commission on the Prevention of Discrimination and Protection (April 1992).

[99] Art 4.

[100] Art 6(e).

The Declaration further maintains that victims should be guaranteed assistance from the criminal justice system (including information that such assistance is available),[101] as well as formal and/or informal procedures providing them with redress, restitution and compensation for wrongs suffered.[102] The document also has provisions protecting victims from unnecessary delay or inconvenience within the criminal justice system[103] and ensuring their concerns are considered. Domestically, the Declaration has closely influenced the development of (usually non-binding) codes of practice in most developed jurisdictions regarding the treatment of victims by the criminal justice system.

In Europe, the 2001 EU Council Framework Decision on the Standing of Victims in Criminal Proceedings now requires all EU states to afford victims basic levels of services and support. The document has spurred a revival of 'rights' language (and associated non-governmental pressure) associated with victims across Europe building upon the various recommendations previously made by the Council of Europe on this matter.[104] The Framework Decision makes similar guarantees to victims as the 1985 UN Declaration, including respect and recognition by criminal justice actors, the right to protection from intimidation and the provision of mediation schemes. Unlike the UN Declaration of 1985, the 2001 Framework Decision defines victims narrowly as those suffering harm 'directly caused by acts or omissions that are in violation of the criminal law of a Member State', which would exclude victims of climate change in virtually all instances. The document has been the subject of extended critique for its narrow conceptions and a new Directive is presently being devised as its replacement.

The above paragraphs notwithstanding, the question of whether such developments afford victims genuine 'rights' is a complex one. In particular, because the mechanisms for the realisation of such 'rights' within individual jurisdictions are usually limited to the internal complaints mechanisms of justice agencies, the *enforceability* of such rights is left open to question. As noted by Jackson.

> One of the problems with putting obligations on criminal justice agencies, however, is that they are unlikely to be taken seriously unless consequences attach to non-compliance.[105]

[101] Art 6(a).

[102] Art 12.

[103] Art 6(e).

[104] In particular, Recommendation (R (83) 7) calling for Member States to develop policies to provide assistance to victims, and to take account of the public's view in forming policies and Recommendation (R (85) 11) on the position of the victim in the framework of criminal law and procedure.

[105] J Jackson, 'Justice for All: Putting Victims at the Heart of Criminal Justice?' (2003) 30 *Journal of Law and Society* 309–26, 139.

Such sentiments can equally be applied to any alleged right to a clean environment, and again begs the question of how those subject to environmental damage brought about through climate change can seek to assert such a right in the international legal order.

The second main option for establishing rights both for victims as a whole and for victims of climate change in particular is to ground such expectations in *existing* rights which have already met with wide acceptance. Here the European Court of Human Rights (ECtHR) has taken the lead in both areas in adapting the Convention to meet such needs. As such, several rulings from the Court now appear to offer victims of crime giving evidence in court a degree of protection from undue intimidation. For example, the cases of *Baegen v Netherlands*[106] and *Doorson v Netherlands*[107] confirm that keeping witnesses (including victim witnesses) anonymous in order to reduce intimidation and enhance protection does not breach a defendant's Article 6 rights under the Convention[108] to a fair trial, provided their evidence can still be effectively challenged. In addition, the case of *Sn v Sweden*[109] confirms that Article 6 does not grant the defence an unlimited right to secure the appearance of witnesses in court. The *Sn* case also maintains that witnesses can give evidence through recorded interviews without breaching Article 6. Doak has gone further to argue that victims might be in a position to argue for a breach of Articles 3 and 8 of the European Convention on Human Rights if they are treated in a degrading manner by the criminal justice system or the state fails to protect their rights to privacy when giving evidence in court.[110] In matters of environmental harm the ECtHR has adopted the Convention in particular to compel states to regulate and actively prevent activities which infringe upon a person's right to (healthy) life and the use of their private property.[111] This approach has the advantage that it does not require consensus on what a good or healthy environment is (as does the less anthropocentric option) merely that the activities in question infringe upon other rights.

Beyond the ECtHR, the 'greening' of existing rights for the victims of environmental harm and the adaptation of such rights for victims as whole is notably less developed. Oda J of the International Court of Justice has referred to the notion of victims having rights in his dissenting opinion in the case of *LaGrand* (*Germany v United States of America*).[112] Likewise in

[106] Series A no 327-B (1995) 77.

[107] *Reports of Judgments and Decisions* 1996-II [1996] 23 EHRR 330.

[108] Convention for the Protection of Human Rights and Fundamental Freedoms, Rome, 4.XI.1950.

[109] (2004) 39 EHRR 13.

[110] J Doak, 'The Victim and the Criminal Process: An Analysis of Recent Trends in Regional and International Tribunals' (2003) 23 *Journal of Legal Studies* 1–32.

[111] See *Taskin v Turkey* [2006] 42 EHRR 50, para 118.

[112] *Order on Provisional Measures*, 1999 ICJ Report 9 at para 9.

the case of *Port Hope Environmental Group v Canada*[113] the UN Human Rights Committee accepted that dumping nuclear waste does raise serious right-to-life issues. Nevertheless, the general lack of development beyond the ECtHR of rights for either group (victims as a whole or those of climate change specifically) makes it difficult to establish a foothold for such concepts in international customary law.

Victim Participation?

So far this section has focused on the recognition of victims of climate change as wronged parties under the umbrella of human rights. Perhaps more significant however is the question of their *participation* in any process in which states are held to account for the harms caused by polluting activities or, in more legal terms, their *locus standi*. The International Law Commission in its draft principles discussed above positively remarked in its studied commentary on the present situation that '[t]he definition of victim is thus linked to the question of standing'.[114] Indeed, legal standing is in fact at the heart of most of the debates concerning victims' rights in general because, in recent history, victims have lacked party status in most domestic criminal justice systems. As a consequence of this legal reality, it is the state which acquires the rights of individual victims, effectively forcing those affected by crimes and other social misfortunes to take a subsidiary role in proceedings. This was famously captured by Nils Christie who argued that the state effectively 'steals' conflicts from their rightful owners, namely the victim and the accused.[115] The same could aptly be said of the victims of climate change being excluded from the international legal order, indeed in this instance the victims' plight is almost entirely taken over by the state because, as already discussed, under international law, it is the state and not the individual that can draw on the international legal order against another state.

It is important to appreciate that a 'right to participate' covers a breadth of possibilities, ranging from the nominal to the much more substantive. Edwards, for instance, has labelled participation 'a comfortably pleasing platitude', which is rhetorically powerful but conceptually abstract.[116] In his discussion, Edwards describes four possible forms of victim participation in criminal justice. The most significant casts victims in the role of *decision-makers*, such that their preferences are sought and applied by the criminal justice system. Less far-reaching would be *consultative*

[113] UNHRC Decision 67/1980.
[114] n 30 above at 137.
[115] N Christie, 'Conflicts as Property' (1977) 17 *British Journal of Criminology* 1–15.
[116] Edwards, above n 88, 973.

participation, where the system seeks out victims' preferences and takes them into account when making decisions. Edwards sees the traditional role of victims in terms of *information provision*, where victims are obliged to provide information required by the system. Finally, under *expressive* participation, victims express whatever information they wish, but with no instrumental impact; here Edwards highlights the danger of victims wrongly believing their participation will actually affect decision-making. A common distinction drawn in these debates is that between 'service rights' and 'procedural rights'. For Ashworth, victim participation should not be allowed to stray beyond service rights into areas of public interest.[117] Ashworth is particularly concerned by victims being afforded the right to influence sentencing (and other decision-making within the process) citing the difficulties of testing victims' claims and taking account of unforeseen effects on victims. The more limited service rights Ashworth has in mind include respectful and sympathetic treatment, support, information, court facilities and compensation from the offender or state, but exclude consultative participation.

For victims of climate change, the 1992 Rio Declaration in one of its most forward-reaching provisions reflects this move towards individual participation in Principle 10, which encourages states to ensure 'public awareness and participation by making information widely available'. It goes on to say that '(e)ffective access to judicial and administrative proceedings, including redress and remedy, shall be provided'. Principle 10 has been adapted by the 1998 UNECE 'Aarhus' Convention on Access to Information, Public Participation in Decision-Making and Access to Justice in Environmental Matters, which requires governments to bring individuals who may be affected into the *decision-making* process when environmental issues are at stake, what Edwards would call *consultative participation*. In terms of holding states to account for the harm caused by climate change what is significant about the Aarhus convention is that, almost uniquely within international law, members of the public are able to refer possible breaches of their rights under the Convention to its Compliance Committee.

The key question from a victimological perspective is whether these provisions under Aarhus are the equivalent of giving victims a level of discretion over what are effectively prosecution decisions (a 'procedural right'), something that has thus-far been vigorously avoided in all jurisdictions, or whether it is merely akin to the complaints mechanisms offered to many other types of victims the world over (a 'service' right). So far there have been few applications made to the Compliance Committee, although it is known that the body cannot issue binding decisions, but rather makes

[117] Ashworth, above n 87 above.

recommendations to the full Meeting of the Parties.[118] This is remarkably similar to the most common mechanism of affording victims 'rights' in most individual jurisdictions, whereby such victims can complain in the event of poor treatment or breach of the relevant codes of practice by the criminal justice system to an independent arbiter or ombudsman, who can then investigate the claims, although usually not enforce any penalty. Hence, the apparent absence of genuinely enforceable rights for individuals concerning the impacts of climate change mirrors the absence of judiciable rights afforded to victims of crime in individual jurisdictions (and by international bodies). In both cases, such rights as are purported to exist for victims lack specific enforcement mechanisms or consequences for those breaching the alleged rights. Thus, while the Aarhus convention offers something of a 'way in' to the international legal order for the individual, albeit through procedural rights only, in practice these Compliance Mechanisms can be subject to many of the same criticisms concerning victim empowerment provisions more generally: there is a lack of real compulsive power on behalf of the Compliance Committee to really address victims' complains and ensure restitution/apologies from perpetrator states is forthcoming.

In sum, this section has demonstrated the application of similar concepts of human rights to both victims as a whole in the victimological literature and specifically to victims of climate change. It has been demonstrated that both groups face the common difficult of a lack of enforcement potential for any alleged rights, calling into question their categorisation as 'rights' at all. Despite such similarities of approach, it is however important to acknowledge that international environmental lawyers and traditional victimologists are in fact approaching human rights from a different perspective. This is because the key difficulty faced by victimologists is the lack of enforcement and redress mechanisms in traditional law for victims of crime. Victimology has made considerable progress towards the *recognition* of such individuals, but lacks the structural basis in law to take the needs of those individuals forward in a substantive way. Conversely, while it has been said that international environmental lawyers face similar difficulties in terms of the enforcement of rights for victims of climate change, this is not because a structure of international redress is not present, but rather because the structure which does exist has not recognised the needs or harm of those *individuals*.

In light of the above discussion, it is argued that the convergence on human rights by international environmental lawyers and by victimologists in fact reflects the structure/agency biases in both disciplines, which were considered earlier. Human rights therefore serves as a useful example of the two disciplines under discussion working in the same language and

[118] Birnie *et al*, above n 86.

exchanging ideas in a manner which takes both forward, albeit in both areas such rights remain generally undeveloped. Again, the case for and interdisciplinary approach seems irresistibly compelling.

CONCLUSION

Climate change is a reality raising significant challenges across the spectrum of social and physical sciences. As such, it seems likely that the solution to the problems caused by climate change must inevitably come from a meeting of minds between scholars and practitioners from diverse fields of inquiry. This chapter has proposed one such collaboration, between criminologists and international environmental lawyers, as offering real conceptual and practical benefits to both fields. Inevitably working together will require compromise and adaptation of traditional ideas in both fields, and indeed it is impossible in a piece of this length to anticipate all such difficulties. Nevertheless, ultimately the fruit of such collaboration will be the furthering of our understanding of how the real harms caused to individuals by climate change (amongst the many other harms so caused) can and should be addressed.

13

Situating Climate Change in (International) Law: A Triptych of Competing Narratives

TAWHIDA AHMED AND DUNCAN FRENCH

triptych: 'a picture or carving... in three compartments side by side' (Oxford English Dictionary)

INTRODUCTION

FOR GOOD OR ill, climate change has come to absorb much of the political and academic attention in the field of international environmental law and governance. Without entering into the debate as to the significance now being paid to climate change vis-a-vis other environmental or 'sustainable development' concerns, such as biodiversity loss or desertification,[1] this chapter seeks to examine the discourse that has emerged on the issue in recent years and, in particular, to consider alternatives to the perception that the multilateral climate change regime[2] is the exclusive vehicle for climate change debates in international law.

[1] This is not to suggest that addressing climate change is not fundamental to appropriately addressing these other issues (see, for instance, A Trouwborst, 'International Nature Conservation Law and the Adaptation of Biodiversity to Climate Change: a Mismatch?' (2009) 21 *Journal of Environmental Law* 419–42, 430: 'climate change is now placing demands on international nature conservation law which are fundamentally different from, and more severe than, the demands for which most conservation treaties were originally negotiated'). However, what is of concern must be any sense that these other issues are increasingly only finding a relevance through their contribution to tackling climate change rather than being important per se.

[2] The 'multilateral climate change regime' is understood here as referring to the 1992 UN Framework Convention on Climate Change (UNFCCC), the 1997 Kyoto Protocol and associated institutions, decisions, processes and legal agreements subsequently negotiated. It must also incorporate an understanding of the political, policy and scientific debates which are part-and-parcel of the climate change regime, thus including both the work of the Intergovernmental Panel on Climate Change (IPCC) and the elaboration of such political texts as the 2009 Copenhagen Accord.

More specifically, the chapter considers three pairs of competing narratives, which highlight not only many of the underlying tensions within the *political* negotiations on climate change—as seen at recent sessions of the Conference of the Parties—but also reflect how *theoretical* enquiries especially in the field of international law have been increasingly framed. These 'pairs' of narratives (which focus upon macro-ideas of law's normativity, context and scope) are not, between themselves, mutually exclusive, though each one does approach the issue of law's response to climate change from a quite distinct perspective. Together, they provide an increasingly more complex—many might add conceptually richer—assessment of international climate change law than one that focuses exclusively, even principally, upon the multilateral paradigm.

What is equally interesting are the competing narratives within each pair, representing apparently opposing viewpoints on how to approach and interpret the role of law in the climate change regime. The three pairs of competing narratives identified are thus (i) 'analytical <-> normative', (ii) 'atomistic <-> integrationist' and (iii) 'localised <-> globalised'. Though discussed in more depth below, it may prove useful to highlight briefly to what each pair refers. The first concerns the extent to which *legal* discussions over climate change are, or should be, purely of a descriptive nature of the positive law or whether the moral-*cum*-political complexity of climate change invariably prompts a more normative search for the 'ought'—as well as the 'is'—of how a proper functioning climate change regime should be designed and operate. This initial pair of competing narratives is thus subsequently referred to in the paper as being concerned with the *normativity* of the legal inquiry on climate change.

The second 'pair' reflects the extent to which climate change should be categorised as a *sui generis* legal topic, considered and analysed separately from both public international law—and, indeed, international environmental law itself—or whether, by so positioning it, climate change becomes a further example of the increasingly fragmented nature of the international legal system. Are we now at a stage that something described as 'climate change law'—something that may have been justifiably dismissed only a few years ago—is no longer a hollow nomenclature but is, in fact, a term that carries substantive meaning? Indeed, numerous university law departments are now offering a bespoke course on climate change, with anecdotal evidence suggesting it is sometimes more popular than the 'general' international environmental law module. A fad? Perhaps. A response to genuine student-demand? Undoubtedly. But how might one view the longer-term impact of this move towards 'climate change law'; what are the consequences of the inevitable narrowing of the law which will ensue and what does a term such as 'climate change law' incorporate—and perhaps more importantly—what does it exclude? Such an approach also raises important questions as to how far such a tight demarcation of the 'law' might fail to give cognisance to the increasing relevance of legal disciplines such as trade, investment, human rights, and peace and security in tackling climate change; what one might refer to as the horizontal 'cross-fertilization' of climate issues.[3] Thus, the second pair of competing narratives can be loosely described as dealing with the 'context' of the climate change regime.

The third 'pair' of narratives reflects the extent to which the climate change regime should be viewed primarily in terms of global governance—viewing the Convention and Protocol as having established an almost monolithic system of treaty provisions, intra-regime decisions and institutional processes—or whether the focus (perhaps even that of international lawyers) should be across a much more diverse array of political and regulatory relationships, including regional and domestic law (and litigation), corporate governance, voluntary codes, and so such a list might go on. In short, climate change may be a global issue but, as an object of legal analysis, why should the climate change regime as developed through the traditional processes of international law provide the only—or even the primary—normative content and point of reference from which to understand the issue? The third pair of competing narratives thus raises the 'scope' of the climate change regime.

It is not the objective of this paper to question the validity of any of these particular approaches; indeed, all have valuable things to contribute to a fuller understanding of the issue. Moreover, as with any topic of legal debate, there is an inevitable movement away from the paradigmatic understanding and towards the articulation of much more critical, circumspect—even, sometimes, just tangential—approaches.[4] In this regard, climate change is no different from international environmental law, and indeed the environmental literature, more generally.[5] However, whether because climate change brings together—arguably in quite a unique way—matters of science (and scientific uncertainty), philosophy, ethics, politics, economics, and law inter alia, or because it does so on an issue of such pressing global importance, the *problematique* of climate change seems a particularly apposite topic on which to consider these trends. Perhaps unsurprisingly, in all the discussions on climate change, little time has been spent considering what it means for climate change to be an *object* of study, in contrast to being a *subject* of both

[3] This is a general issue for international environmental law, on which P Birnie, A Boyle and C Redgwell, *International Law and the Environment*, 3rd edn (Oxford, Oxford University Press, 2009) 3: '"international environmental law" is of course different from international human-rights law, the law of the sea, natural resources law, or international economic law, inter alia, but there are significant overlaps and interactions with these categories, and the categorisation *is in some cases a matter only of choice and perspective*'. Emphasis added.

[4] For a very good example of this, see L Godden, 'Death, Desire, Modernity and Redemption: Climate Change and Public International Environmental Law' (2009) 10 *Melbourne Journal of International Law* 543–78. If the title itself does not give something of the nature of the approach away, the author notes 'the article adopts a style that is meant to challenge the view that there is a sovereign, authoritative meaning to text (and law) that is readily discernible!'

[5] For instance, for an excellent collection of critical insights into environmental law, see B Richardson and S Wood (eds), *Environmental Law for Sustainability* (Oxford, Hart, 2006).

academic inquiry and intense political negotiation. Thus, this paper makes an initial attempt to focus upon the former, though this cannot be divorced from the substance and politics of climate change, which inevitably—and rightly—has dominated the literature, and comprises the latter.

To assist in this analysis, these narratives are explored through a critical questioning of what two distinctive, yet inter-related, approaches to climate change might bring to the debate; first, the use of human rights norms and language and second, a reliance on themes of fairness and justice. Though both approaches are similar, to the extent that they seek to embed climate change within a broader context—a context replete with terms such as 'human-centred', 'climate justice' and 'solidarity'—their premise and framework of reference are however quite different. The former is much more firmly embedded within established legal norms—though its application to the particularities of climate change is, of course, contentious—whereas the latter touches upon issues, such as distributional equity, on which the law is only beginning to find coherent expression.

In conclusion, it is suggested that studying the climate change regime provides a very large canvas on which to sketch very diverse, if not apparently contradictory, normative and political suppositions. Though such an exploration of ideas is a positive characteristic of the evolution of a new field of study and research—and certainly is central to the academic endeavour—might it not also indicate, rather more problematically, the continued absence of a meaningful consensus as to how to best respond meaningfully to the scale and complexity of climate change?

CLIMATE CHANGE: UNIVERSALIST PARADIGM, COMPETING NARRATIVES

Climate change can easily be described in universalist terms, not only because of the nature and 'reach' of the problem but also because of the global scale of action required to tackle the myriad of issues that go to comprise it. As the following quotations indicate, the rhetoric is compelling:

> change in the Earth's climate and its adverse effects are a common concern of humankind.[6]
>
> We recognize that climate change is a serious and long-term challenge that has the potential to affect every part of the globe.[7]
>
> Unmitigated climate change would, in the long term, be *likely* to exceed the capacity of natural, managed and human systems to adapt.[8]

[6] UNFCCC, preamble.

[7] World Summit Outcome 2005 (UN Doc A/Res/60/1 (24 October 2005)), para 51.

[8] Intergovernmental Panel on Climate Change (IPCC), 'Fourth Assessment Report', *Synthesis Report* (2007).

We underline that climate change is one of the greatest challenges of our time.[9]

Thus, climate change is perceived, both geographically ('every part of the globe') and morally ('one of the greatest challenges of our time'), as universal.[10] Moreover, climate change is increasingly universalist in terms of the areas of economic, social and political life that are affected by the challenges that its represents.[11] Unsurprisingly, this view of climate change as a universal problem, requiring universal action—in part, because of the all-encompassing moral questions that it poses of humanity as a whole—invariably pushes towards a view of the multilateral climate change regime as the principal, if not exclusive, normative expression of the political need to act. In short, global governance is, by definition, a proxy for the inevitability of universalism.[12] As the 2005 World Summit Outcome perhaps rightly, but also somewhat idealistically, noted '[t]he Convention is the appropriate framework for addressing future action on climate change at the global level'.[13] And though there is, we think, an interesting question here over how far there is an epistemological parasitic relationship between the 'issue' of climate change and the international legal and scientific response thereto and, more particularly, how the latter has seemingly

[9] 2009 Copenhagen Accord, available at unfccc.int/files/meetings/cop_15/application/pdf/cop15_cph_auv.pdf, last accessed: March 2010, para 1.

[10] For instance, see S Caney, 'Cosmopolitan Justice, Rights and Global Climate Change' (2006) 19 *Canadian Journal of Law and Jurisprudence* 255–78; and J Ford, 'The Role of Culture in Climate Change Policy Making: Appealing to Universal Motivators to Address a Universal Crisis' in A Carvalho (ed), *Communicating Climate Change: Discourses, Mediations and Perceptions* (Braga, Centro de Estudos de Comunicação e Sociedade, Universidade do Minho, 2008) available at: www.lasics.uminho.pt/ojs/index.php/climate_change 73: 'Climate change poses the first universal crisis of our planet. The natural world recognizes none of man's arbitrary national borders, trade pact zones, or gated communities and affects all humans in an unpredictable manner. It is an urgent and universal crisis that demands an urgent and universal policy. But is universal policy possible in a world divided by cultural differences?'.

[11] Undoubtedly, law has yet to reflect the comprehensive change required by climate change, especially at the level of the societal and the individual. Equally, on the economic front, the full-scale of the change required has still to be fully grasped. Nevertheless, some of the scope of this change is perhaps captured by article 2 of the 1997 Kyoto Protocol which requires Annex I Parties (developed States and States with economies in transition) to '[i]mplement and/or further elaborate policies and measures' on such matters as energy efficiency, forestry, agriculture, 'new and renewable forms of energy, of carbon dioxide sequestration technologies and of advanced and innovative environmentally sound technologies', 'market imperfections, fiscal incentives, tax and duty exemptions and subsidies in all greenhouse gas emitting sectors', transport, waste management, and the 'production, transport and distribution of energy'. Later on in the same provision, there are references to the carbon implications of aviation and international maritime transport, thus extending the reach of the climate change regime to key drivers—and conduits—of the global economy. In short, action to address climate change touches upon virtually all aspects of the agricultural, industrial and post-industrial economy.

[12] See, for instance, JI Charney, 'Universal International Law' (1993) 87 *American Journal of International Law* 529: 'The international community of the late twentieth century faces an expanding need to develop universal norms to address global concerns. Perhaps one of the most salient of these concerns is to protect the earth's environment'.

[13] World Summit Outcome (2005), above n 7, para 51.

come to 'own' the former,[14] it is nevertheless reasonably apparent that the multilateral climate change regime has become, by instinctive default, the paradigmatic understanding of the climate change issue.

Moreover, though the express differentiation of commitments within the global regime, particularly between developed and developing countries, might be considered as militating against such universalism, this is not necessarily the case. Indeed, we would suggest that the very existence of such differentiation presupposes the a priori commonality of a joint endeavour. As noted above, the preamble to the 1992 Convention very resolutely states, climate change is a 'common concern of humankind'. Though such language arguably lacks legal precision, it is more than adequate to provide normative authority—if not also moral justification—for the multilateral regulatory paradigm. As Brunnée, for instance, has noted,

> common concern regimes focus on equitable sharing of the burdens of cooperation and problem solving. These considerations are encapsulated in the concept of common but differentiated responsibilities, which may fairly be described as the flipside of the concept of common concern.[15]

Thus, as a rhetorical device—with increasingly normative implications—common concern is perfectly synergistic with the universal appeal of climate change.[16] Moreover, though this paper cannot engage with the furore surrounding the attempt by a few key countries to persuade the treaty membership as a whole to endorse the highly political Copenhagen Accord in the concluding hours of the 2009 session of the Conference of the Parties,[17] an implicit criticism raised by this attempt was the inevitable sense that this was destructive of the global nature of the multilateral regime.[18] As noted above, if multilateralism is a proxy

[14] *Cf* B Wynn, 'Scientific Knowledge and the Global Environment' in M Redclift and T Benton (eds), *Social Theory and the Global Environment* (London, Routledge, 1994) 188: 'The sociology of global environmental science...would imply the reopening of explicit and diverse negotiations of pluralist epistemological commitments within a more open-textured, culturally differentiated and socially permissible "scientific" realm...This focus on the implicit cultural framing of scientific knowledge does not mean that such knowledge would be debunked or denied authority. Rather the conditions of validity would be critically explored, and the tacit social and moral commitments of knowledge exposed for debate and negotiation'.

[15] J Brunnée, 'Common Areas, Common Heritage, and Common Concern' in D Bodansky, J Brunnée and E Hey (eds), *The Oxford Handbook of International Environmental Law* (Oxford, Oxford University Press, 2007) 566.

[16] Brunnée, however, makes the important point that one needs to be careful what it is one is, in fact, saying is of common concern—'it is not the atmosphere or even the "climate", as such, that is the common concern but rather the '*change* in the Earth's climate and its *adverse effects*" (preamble)' (ibid, 565).

[17] For general analysis of the Copenhagen Accord and the negotiations surrounding it, see D Bodansky, 'The Copenhagen Climate Change Conference: A Post-Mortem', available at papers.ssrn.com/sol3/papers.cfm?abstract_id=1553167 (last accessed: March 2010).

[18] L Rajamani, *Copenhagen Accord: Neither Fish nor Fowl* (February, 2010) www.cprindia.org/morepapers.php?s=158, last accessed: March 2010: 'The Copenhagen Accord was reached

for universalism, anything that might weaken that multilateralism is inherently distrusted especially by the weaker members—some might say 'silent majority'—of the international community as moving away from the generally accepted view of how climate change as a multi-stranded problem should be both tackled and governed.

None of this would matter greatly—climate change is, indeed, a global issue!—but for the fact that a paradigm has the tendency to absorb and/or subvert alternative perspectives. And in de facto restricting attention to the multilateral regime, the risk is that this ignores the broader—undoubtedly more complex, but also conceptually enriching—normative, political and social framework that has evolved around climate change. Thus, in seeking, in part, to move away from an approach exclusively dominated by a singular approach or perspective, this paper will now move on to consider the three pairs of competing narratives, which, as noted above, broaden the normativity, context and scope of our approaches to the legal understanding of climate change. This is not to suggest that the approaches identified are somehow exhaustive—far from it—but rather that they highlight increasingly diverse perspectives and issues, which might otherwise not be considered if the topic is viewed from one particular perspective.

NORMATIVITY

The first pair of narratives identified is 'analytical <-> normative' and reflects the tension, which is admittedly visible in many areas of international law, between *lex lata* and *de lege ferenda*, between analysing the law as it currently exists and seeking to explore, to ponder—even to hope—for the law as it might be.[19] In short, this first pair of narratives highlights the extent to which legal analysis of climate change often strays beyond a

among 29 states, including all major emitters and economies, as well as those representing the most vulnerable and least developed; The Conference of Parties (CoP) neither authorized the formation of this group to negotiate the accord, nor was it kept abreast of the negotiations as they evolved. As this occurred after ten days of repeated procedural irregularities and ill-considered initiatives by the Danish presidency, patience and confidence were wearing thin. Therefore, when the accord was presented to the CoP for adoption late on December 18, it was categorically rejected by, among others, Bolivia, Cuba, Nicaragua, Sudan, Venezuela and Tuvalu. They did so both because of the manifest procedural irregularities in the negotiation of this accord as well as the substantive weaknesses they perceived in it. As CoP decisions require consensus (not unanimity) for adoption, the Conference of Parties, in a night marked by unparalleled histrionics and presidential ineptitude, could only resolve to 'take[s] note' of the Copenhagen Accord'.

[19] *Cf* P Weil, 'Towards Relative Normativity in International Law?' (1983) 77 *American Journal of International Law* 421: 'the distinction between lex lata and lex ferenda must be maintained with no abatement of either its scope or its rigor...Without this positivistic approach, the neutrality so essential to international law qua coordinator between equal, but disparate, entities would remain in continual jeopardy'.

doctrinal assessment of the current legal framework to engage with broader questions of idealised purpose and design. And while the extensive body of rules and procedures that now comprise the climate change regime are more than sufficient to provide both positivist (and indeed more critical[20]) scholars with a rich analytical feast, the universal nature of the issue also permits a significant degree of latitude in considering what the legal response to climate change might—or *ought* to—be. As with contemporary discussions over the reform of the UN, the issue of climate change and the multilateral regime put in place to tackle it, encourages much ruminating on both the nature of global governance and, more widely, the present state of the international community. On an issue for which it is difficult not to have a view—or to keep separate the personal from the academic—it becomes increasingly hard to divorce one's opinion as to how to respond to climate change from an analytical assessment of the current regime. Thus, it seems that these perspectives are perhaps less opposite understandings on climate change (ie either one chooses whether to approach the issue as a matter of positive law or as a question of normative development) and rather that they converge—be it implicitly, even subconsciously—into the same stance and understanding. In other words, in assessing the current state of the law one risks also seeing the inclusion of one's own view as to the direction in which the law should be heading.

The human rights approach to climate change is a pertinent demonstration of this conflation of views. The continued survival and relevance of human rights discourse depends on making the analytical claim, in other words, on demonstrating that the existing corpus of international human rights obligations already imposes duties in relation to the environment and at times specifically climate change, as implemented through litigation, international supervision, the diplomatic language of international discourse and domestic law.[21]

However, a substantive 'human right to the environment does not yet exist in international law and... approaches towards the creation of such a right have largely halted.'[22] But, human rights are a powerful tool for the development of new ideas, and necessarily and explicitly take on the challenge of

[20] See K Mickelson, 'Critical Approaches' in Bodansky, Brunnée and Hey (eds), above n 15, 289 who notes that '[c]ritical approaches have doubtless have some influence [in the international environmental law literature], but they have largely remained on the periphery of scholarly engagement'. Indeed, as an important point of methodology, she remarks that such critical approaches 'do not limit themselves to critique. Instead, they provide creative and often inspiring ways of reconceptualizing our responses to environmental problems'.

[21] See eg before the ECtHR, *Lopez Ostra v Spain* [1994] 20 EHRR 277; *Oneryildiz v Turkey* [2004] 39 EHRR 12. *Budayeva v Russia* [2008] ECHR 15339/02. Before the African Commission on Human Rights, see the Ogoni case—SERA Centre and *CESR Centre v Nigeria* (2003) 10 IHRR 282. For further discussion, see chapter 7 (Obokata) and chapter 12 (Hall).

[22] O Pedersen, 'European Environmental Human Rights and Environmental Rights: A Long Time Coming?' Electronic copy available at: ssrn.com/abstract=1122289 (last accessed: January 2011), 1–50, 3.

reform and change.[23] Thus adding to the identification of what the law is, human rights policy sets out a complete picture of the (potential) relevance of human rights to climate change. This primarily involves the imaginative use of established human rights, but may also call for the introduction of new human rights. Thus, both the Office of the High Commissioner for Human Rights (OHCHR) and Human Rights Council have asserted that climate change potentially affects the 'full' range of human rights.[24] In a similar vein, General Comments from the Human Rights Committee and the Committee on the Rights of the Child have emphasised the interdependency of the rights to life, health and environment.[25] Human rights talk is thereby introduced in many disparate areas of human activity including migration or displacement,[26] transnational corporations,[27] or global peace and security.[28] Such pronouncements are made irrespective of whether such duties might actually be accepted within the international community. Thus, even if international documents cannot always claim an existing duty or highlight the existence of a relevant enforcement mechanism, the implicit message is that such duties ought to exist or, at least, be more strategically utilised. The obvious criticism is of course that these claims are idealist. The OHCHR itself acknowledges this. Thus, in the 2009 OHCHR report on climate change, there is recognition that claims that climate change breach human rights do not always translate into readily-identifiable duties and, in

[23] O Pedersen, 'Climate Change and Human Rights: Amicable or Arrested Development?' (2010) 1 *Journal of Human Rights and The Environment* 236–51, 238; S Trully, 'The Contribution of Human Rights as an Additional Perspective on Climate Change Impacts Within the Pacific' (2007) 5 *New Zealand Journal of Public International Law* 170; S Trully, 'Like Oil and Water: A Sceptical Appraisal of Climate Change and Human Rights' (2009) 15 *Australian International Law Journal*; A Sinden, 'Climate Change and Human Rights' (2007) 27 *Journal of Land, Resources and Environmental Law* 255.

[24] Report of the Office of the United Nations High Commissioner for Human Rights on the Relationship between Climate Change and Human Rights, UN Doc A/HRC/10/61, para 20; Human Rights Council Resolution 7/23 (2008) preamble. See also the EU Charter on Fundamental Rights, Art 37 which states that environmental protection is to be integrated across all EU policies.

[25] Human Rights Committee, General Comment 6, para 5; Human Rights Committee, General Comment 28, para 10; Committee on the Rights of the Child, General Comment 7, para 10; Committee on Economic, Social and Cultural Rights, General Comment 12, para 4; Committee on Economic, Social and Cultural Rights, General Comment 14, para 4.

[26] IPCC, *First Assessment Report (AR1)* (1990); AR4.

[27] See the Ogoni case—*SERA Centre & CESR Centre v Nigeria* (2003) 10 IHRR 282.

[28] Eg in 2007, the Security Council held a day-long debate on the impact of climate change on peace and security, cited in Report of the Office of the United Nations High commissioner for Human Rights on the Relationship between Climate Change and Human Rights, UN Doc A/HRC/10/61, para 61; UN Sub-Commission, 'Commission on Prevention of Discrimination and Protection of Minorities, 'Review of Further Developments in Fields With Which the Sub-Commission Has Been Concerned, Human Rights and the Environment.' UN Doc E/CN.4/Sub.2/1994/9 (July 6 1994) (prepared by Special Rapporteur Fatma Zohra Ksentini); see also J Nickel, 'The Human Right to a Safe Environment: Philosophical Perspectives on Its Scope And Justification' (2003) 18 *Yale Journal of International Law* 281.

any event, these duties are not, in all cases, enforceable.[29] More provocatively, Bodansky notes:

> Legally, climate change no more violates human rights violation than does a hurricane, earthquake, volcanic eruption, or meteor impact.[30]

Bodansky aims here to distinguish between desires and actual legal rules. Although human rights discourse pushes for the recognition of profound links between the impact of climate change and breaches of human rights law, the current state of international human rights law is ill-equipped to bring such desires to fruition and to tackle such levels of complexity. Very few impacts of climate change are actually enforceable by human rights law, especially against potentially such a wide array of actors, including non-State actors. Nonetheless, the language of human rights is seen as having immense symbolic power. As Mary Robinson states,

> the human rights framework gives us the legal and normative grounds for empowering the poor to seek redress.[31]

Most recently, this view is slowly being accepted by the UNFCCC itself, demonstrating an acceptance of the role of human rights by the central international institution on climate change.[32]

The tendency to move between the 'is' and 'ought' is not unique to human rights. It also arises if one seeks to understand climate change from a fairness perspective, which must invariably blur the law as it stands and the law as it ought to be. The current law may be a benchmark for doctrinal comment and criticism, but by definition it is impossible to conduct a discussion on its fairness alone through a positivist analysis of the law. A fairness discourse thus seeks to range beyond a textual analysis of international environmental law to understand its moral, distributional and procedural elements.[33]

[29] Report of the Office of the United Nations High commissioner for Human Rights on the Relationship between Climate Change and Human Rights, UN Doc A/HRC/10/61, para 70.

[30] D Bodansky, 'Climate Change and Human Rights: Unpacking the Issues', available at papers.ssrn.com/abstract=1581555, last accessed: April 2010; see also J Knox, 'Climate Change and Human Rights Law' (2009) 50 *Virginia Journal of International Law* 163.

[31] M Robinson, *Climate Change and Justice*, Barbara Ward Lecture, London, Chatham House, 11 December 2006.

[32] Draft Decision 1/CP.16 (Cancun(2010)), Outcome of the work of the Ad Hoc Working Group on long-term Cooperative Action under the Convention, preamble recital 7; and para 8: '*Emphasizes* that Parties should, in all climate change-related actions, fully respect human rights'.

[33] As an aside, it should be noted that though ideas such as 'fairness', 'justice' and 'equity' are used reasonably interchangeably throughout this paper, it is recognised that there are arguably important differences of meaning between them and, more specifically, that 'equity', in particular, has been used frequently in a much more defined, and circumscribed, manner in international law. However, though 'equity' does appear to be more of a 'legal' term-of-art than perhaps 'justice' or 'fairness', this apparent textual certainty should be understood within the broader context that equity will still often afford a significant degree of discretion to decision-makers. Moreover, it is worth reminding oneself that the incorporation of overtly socio-economic and economic justice considerations within the decision-making process is not per se excluded by the supposedly more 'legal' idea of equity, but that it is very much dependent upon

As Ebbesson remarks, '[e]nvironmental laws and policies are predominantly goal-oriented... Yet, environmental law also involves priorities, conflicts and clashes of interests—and concerns for justice and fairness'.[34] This concern for fairness incorporates both a substantive and a procedural element. Three aspects seem especially apposite in this regard—the climate change regime being perfectly situated to reflect the entirety of this debate. First, a concern for fairness raises whether, and how far, one should be concerned with correcting historic inequalities in the use of the Earth's natural resources and, in the case of climate change, also its sink capacity (what one might refer to as the search for 'corrective justice'). Second, a concern for fairness raises whether, and how far, one should be concerned with seeking the negotiation of fair—or again, just—prospective outcomes (the search for 'distributive justice'). And third, a concern for fairness raises the need for—the imperative of—'procedural' legitimacy. As Hey notes in regard to global water law, but the comment seems of broader application,

> procedural fairness needs to be enhanced... by introducing a degree of formalism, associated with the rule of law, in decision-making procedures, both to ensure that the outcomes are fair in terms of distributive justice at the level of both states and individuals, but also to enhance, what Lon Fuller found distinguishes law from other normative systems, good process.[35]

However, while these three conceptions of justice are closely linked, they are neither synonymous nor necessarily mutually inclusive—one would hope for the synergy that Hey propounds between substance and procedure, but this is not necessarily always the case.[36] Nevertheless, each aspect is important, and each brings a different facet of fairness to international environmental law generally, and climate change in particular.[37]

More generally, some might suggest that this conflation of law as is and the law as it might be is intrinsic to much international legal scholarship. Koskenniemi has, for instance, had occasion to remark that international

the subject-matter at issue and, more importantly, the way in which the recourse to equity on the topic has been legally and politically framed.

[34] J Ebbesson, 'Introduction: Dimensions of Justice in Environmental Law' in J Ebbesson and P Okowa (eds), *Environmental Law and Justice in Context* (Cambridge, Cambridge University Press, 2009) 1.

[35] E Hey, 'Distributive Justice and Procedural Fairness in Global Water Law' in ibid, 369.

[36] Indeed, for some, procedure is of less relevance. C Okereke, *Global Justice and Neoliberal Environmental Governance: Ethics, Sustainable Development and International Co-Operation* (London, Routledge, 2008) 35: 'the stuff of any theory of justice resides in the principle of distribution it recommends, and questions about procedure are relevant only as a means to establish the morality of the outcome or combination of outcomes envisaged'.

[37] See Brunnée, above n 15, 329: 'Global climate change is one issue area in which states must face up to what Henry Shue has termed the 'unavoidability of justice'. Some important steps have been taken. Notably, the concept of common but differentiated responsibilities sketches the parameters of a debate *about* global climate justice. However, it does not currently constitute a genuine principle *of* global justice'.

law 'exists as a promise of justice'.[38] True, that he goes on to say that such justice 'cannot be enumerated in substantive values, interests or objectives'[39] but nevertheless there is an aspect to international law which is 'the announcement of something that remains eternally postponed'.[40] Whether one agrees with this view of the international legal system or not, there is little to disguise the fact that much of the environmental law literature—as with human rights—incorporates an assumption as to the inherent value of its own internal objectives; it is not easy for someone to be seen to be disagreeing as to the goal of environmental protection. Of course, one can dispute the detail, but the general principle and desired outcome is usually taken as unquestioned, and being unquestionable.

So why this tendency—admittedly far from being on every issue by every commentator—to fail to demarcate the present from the future; the current from the desired; the positive from the normative? In short, it would seem to us that what is particularly significant as regards the climate change regime is the connection that is made to the broader international system. Not only can the climate change regime be seen as a barometer of the current state of general international law and politics,[41] but it is also often treated as a useful template for speculating as to where we—as an international community—might all go from here in dealing with global problems as a *global* society. Just as with the ongoing discussions on the future of the United Nations, climate change's inherent universalism attracts a particular form of scholarship. In a world of increased global stresses, where sovereignty and the role of the State will be increasingly qualified by the overarching priorities of the international community, the climate change regime provides a point of engagement for both analysing the present content of the law and discerning its normative prospect. And in doing so, the divide between them can be easily missed. While not criticising this—it almost seems intrinsic to the international lawyers' DNA that they envisage a better, more just, world—what perhaps we should do is to recognise that such a convergence does, in fact, happen, perhaps more often than we

[38] M Koskenniemi, 'What Is International Law For?' in M Evans (ed), *International Law*, 3rd edn (Oxford, Oxford University Press, 2010) 52.

[39] Ibid.

[40] Ibid 78.

[41] For instance, see Bodansky, above n 17, 10: 'The Copenhagen Conference is also notable, from a process standpoint, for several shifts in the negotiating dynamic. China was much more assertive, reflecting its emergence as a global power, at one point thumbing its nose at the United States by sending a mid-level official to negotiate with President Obama; conversely, the EU played a less central role. The fractures within the developing country negotiating bloc (the so-called Group of 77) were more evident than ever; indeed, during the final session, one small island state negotiator publicly blamed the big developing countries for the failure to make more progress. Finally, the willingness and ability of a small group of countries that had previously played little role in the negotiations to block adoption of the Copenhagen Accord showed the absurdity of the consensus decisionmaking rule'.

realise, and to more clearly demarcate the one from the other. Else we risk what Weil warned against:

> Without higher moral 'values', international law is but a soulless contrivance... this will to transcend the traditional international society made up of juxtaposed egoisms, and to forge an international community animated by the quest for the 'common good' and common 'values', is all the more precious... Vigilance, however, is imperative... the waning of... positivity in favour of ill-defined values might well destabilize the whole international normative system and turn it into an instrument that can no longer serve its purpose.[42]

CONTEXT

The second pair of narratives is 'atomistic <-> integrationist' and reflects the extent to which the climate change regime should be considered separately, or in conjunction with, other matters of international concern, including trade, investment, development, as well as peace and security, and human rights.

Moreover, this capacity to consider climate change either in isolation or expansively can be viewed at a number of different levels; at the level of the organisational (ie as a self-contained regime or, alternatively, as part of other attempts at global institution-building),[43] at the level of international norms (ie as a discrete body of rules or, alternatively, to consider how these rules are affected by competing norms, for instance those under the remit of the World Trade Organization),[44] or at the level of policy and the political (most prominently whether climate change is a discrete environmental problem or is symptomatic of global un-sustainable development, more generally).[45] Moreover, assessing whether climate change is best viewed through uni- or varifocal lens also forces us to question our own perspectives and view-points on the issues.

But even if one's primary perspective is international environmental law—and not, say, domestic environmental regulation or the law relating to financial transactions to name but two areas which are increasingly affected by, and affecting, climate change law and policies on which, partially, see below—there is still the risk that over-concentration upon the multilateral

[42] Weil, above n 19, 423.

[43] H van Asselt, F Sindico and M Mehling, 'Global Climate Change and the Fragmentation of International Law' (2008) 30 *Law and Policy* 423–49.

[44] For instance, see J Werksman, K Baumert and N Dubash, 'Will International Investment Rules Obstruct Climate Protection Policies? An Examination of the Clean Development Mechanism' (2003) 3 *International Environmental Agreements: Politics, Law and Economics* 59–86.

[45] U Luterbacher and D Sprinz (eds), *International Relations and Global Climate Change* (Massachusetts, MIT Press, 2001).

environmental regime is misleading. Focusing *just* upon the multilateral regime risks exactly the type of fragmentation in the normative order, which the International Law Commission has recently had recourse to consider. The following extract from the ILC's 2006 report on the difficulties arising from the diversification and expansion of international law, we think, is telling:

> The problem, as lawyers have seen it, is that such specialized law-making and institution-building tends to take place with *relative ignorance of legislative and institutional activities in the adjoining fields and of the general principles and practices of international law*. The result is conflicts between rules or rule-systems, deviating institutional practices and, possibly, the loss of an overall perspective on the law.[46]

With particular reference to climate change, 'adjoining fields' would include, but not be limited to, the extent to which the rules of the World Trade Organization and those contained in bilateral investment treaties can be used creatively to support, rather than preclude or restrict, climate change policies and actions, the use of human rights mechanisms to raise—and enforce—the rights implications of climate change,[47] the potential reliance on the general rules on State responsibility to hold States to account for climate change damage (including sea level rise),[48] as well as the spectre of utilising the UN organs, most notably the mandatory powers of the Security Council,[49] to bring climate change ever deeper within the remit of international peace and security.

Of particular interest is the necessity for human rights to mainstream its own normative framework and values into other areas of law and policy. Human rights law and discourse has taken a very integrationist approach to climate change; seeking to apply human rights pervasively to climate change, and vice versa.[50] Beyond seeking to apply human rights to apparently 'non-rights' issues—as how mainstreaming has been traditionally understood—such mainstreaming has also materialised in other forms,

[46] *Fragmentation of International Law: Difficulties Arising from the Diversification and Expansion of International Law*, Report of the Study Group of the International Law Commission, A/CN.4/L.682, 13 April 2006, 11. Emphasis added.

[47] Including both judicial mechanisms (eg European Court of Human Rights) and non-judicial mechanisms, eg Report of the Office of the United Nations High commissioner for Human Rights on the Relationship between Climate Change and Human Rights, UN Doc A/HRC/10/61.

[48] R Verheyen, *Climate Change Damage and International Law: Prevention Duties and State Responsibility* (Leiden, Brill, 2005).

[49] In 2007, the Security Council held its first ever session on the security impacts of climate change, www.un.org/News/Press/docs/2007/sc9000.doc.htm.

[50] Report of the Office of the United Nations High commissioner for Human Rights on the Relationship between Climate Change and Human Rights, UN Doc A/HRC/10/61, paras 33, 39 and 40. See also J Knox, 'Linking Human Rights and Climate Change at the United Nations' (2009) 33 *Harvard Environmental Law Review* 477.

including in increasing references to the second and third generation of human rights across international and domestic tribunals (which is especially significant in the area of climate change) and the growing willingness of international bodies to cross-reference the environment-friendly decisions of other international or domestic bodies, where their own precedents fail to provide the necessary support for the broad interpretations they now seek.[51] Mainstreaming arguably represents the single most influential role human rights can play in the climate change debate. Indeed a specialist human rights approach to climate change, for example through the introduction of an entirely new right, is unlikely in the current political environment, given the more general shortcomings of the international human rights regime as well as an increasing weariness to elaborate 'new' rights. Though a human rights litigation approach is currently likely to provide only incremental benefit in responding comprehensively to climate change concerns, nonetheless the more general attempt to mainstream human rights into an issue that many states might otherwise prefer to remain within the purview of diplomatic negotiations demonstrates the importance of seeking a broader context.[52]

As regards the application of a fairness discourse into the climate change debate, context too would seem to be all-important. A fairness discourse would seem to require the international community to focus on the allocation of resources and the distributional implications thereof in a holistic, or expansive, manner recognising the artifice of valuing the fairness of one regime in the midst of global injustice, more generally. It is therefore no wonder that the preamble to the 1992 Climate Change Convention states that responses to climate change should be coordinated with social and economic development in an integrated manner with a view to avoiding adverse impacts on the latter, taking into full account the legitimate priority needs of developing countries for the achievement of sustained economic growth and the eradication of poverty.

No one would suggest that responding to climate change will resolve the world's ills, but nor should it operate either in ignorance of them or as a mechanism for their further exacerbation.

From these general themes, a particular—and increasingly robust—debate is now taking place around climate change and fairness. A recent issue of the *Melbourne Journal of International Law* was, for instance,

[51] Eg the UN Report on climate change refers to developments within the African Union and American Convention on Human Rights and the Council of Europe. See Report of the Office of the United Nations High commissioner for Human Rights on the Relationship between Climate Change and Human Rights, UN Doc A/HRC/10/61, paras 4, 17, 54, 74, 106.

[52] This is not to ignore the limitations of a human rights approach. See O Pedersen, 'Climate Change and Human Rights: Amicable or Arrested Development?' (2010) 1 (2) *Journal of Human Rights and the Environment* 236–51, 250, who argues inter alia that human rights reduces the climate change debate to the question of 'violators' and 'victims'.

largely given over to the issue; with the editors noting in the foreword that '[i]n focusing on 'climate justice', the symposium places questions of global equity and distributional justice at the core of international debates around climate change mitigation and adaptation'.[53] Of particular significance in this regard was the comment by one of the contributors, Jeremy Baskin in his article, who noted that '[b]y "climate justice" I mean something both beyond and different from the notion of climate law. Climate law may not always be justice'.[54] At the end of his piece, he then says this, worth quoting in full:

> It is generally regarded as naïve, when considering international relations, to focus on justice, or to emphasise right over might. In this case—perhaps uniquely—even the powerful need a genuinely global solution, and agreement cannot be achieved without an engagement with justice. We have to hope that the pattern of millennia can be broken. In this instance, might *needs* right. Climate justice is necessary and we can only hope it is possible. It needs to be higher up the policy agenda than it is currently.[55]

Thus, as a second perspective by which to understand climate change—after a human rights approach—fairness is both attractive as a means of providing an all-engaging understanding of an issue such as climate change while being sufficiently grounded in both theory and political reality to provide (admittedly highly disputed) guidance and strategic tools to States and other key negotiating actors. Moreover, the very fact that fairness and human rights are value-laden concepts, which may be seen to provide an overarching meta-frame—'constitutional' seems too strong in this regard—for climate change, requires us to go beyond the multilateral paradigm as the exclusive approach to climate change. In particular, fairness, as we have already seen, reminds us to take into consideration the political, socio-economic and various other contexts in which climate change operates. A fairness approach also argues for global redistribution. So we need to determine more than simply what our goals are; we need to know how to achieve them on a fair basis. Just as with human rights, a fairness discourse relies upon a mainstreaming or integrationist perspective to climate change to achieve this.

There is, however, another sense in which the second pair of narratives (focused on the context of climate change) operates and it occurs within the confines of the climate change regime itself. The climate change regime has become broad and detailed, dealing with issues as wide-ranging as emissions trading, project-based clean energy, the conservation of forests

[53] 'Foreword' (2009) 10 *Melbourne Journal of International Law* i.
[54] J Baskin, 'The Impossible Necessity of Climate Justice?' (2009) 10 *Melbourne Journal of International Law* 425.
[55] Ibid 438.

for reducing carbon emissions, improving carbon capture and storage, financial and technical assistance from North to South, inter alia, as well as 'procedural' matters of implementation and compliance, to name but a few. Thus, this competing narrative of 'atomistic <-> integrationist' reminds us of the benefits and risks of over-specialisation within the regime itself. What do we lose in both perspective and knowledge if we isolate for specific attention and individual inquiry components of the wider regime? On the other hand, what can truly be gained from remaining at the generic, and the increasingly abstract, level of the regime as a whole? It must be said, of course, that studying individual elements per se is not problematic, though it may become so if such study occurs in isolation and with little—or no—regard for the regime in which such elements form part, never mind, more broadly, of the legal system(s)[56] from which the regime itself secures overall normative authority. Such fragmented analysis presents, we believe, a risk of selective appreciation of the complexity of not only the legal relations—both public and private—that increasingly surround climate change, but also the wider political framework in which they occur. Space precludes a detailed analysis of even a representative example of the difficulties that this over-specialism might cause, though perhaps the following rhetorical question captures something of the issue:

> [c]an we ignore the mechanisms proposed to address climate change—whether market-based regulation (such as emissions trading) or more traditional mandate-type regulation? Should we be neutral about the mechanisms and focus on the ends? ... Similarly in relation to technology, does climate justice require environmentalists to rethink attitudes to nuclear power, "clean coal" and high-tech proposals.[57]

Ignoring the specifics of each of these questions, what they also reveal is the over-arching link that persists between the objectives of, and tensions within, the broader regime and the more detailed aspects of some of the more contentious aspects of implementation. Thus, in considering the *context* of the multilateral regime, both between the regime vis-a-vis the rest of the international legal system and between the regime vis-a-vis individual components therein, the spectrum of how far one should view the regime as an integrative part of something larger not only raises the issue of the appropriate framework for legal analysis but also important methodological questions.

[56] One noticeable aspect of the Kyoto Protocol has been the engagement of business within the flexible mechanisms, which almost by definition raises the prospect of the applicable law not just being international environmental law (between States governed) but also domestic regulation and private contracts. Generally, see D Freestone and C Streck (eds), *Legal Aspects of Implementing the Kyoto Protocol Mechanisms* (Oxford, Oxford University Press, 2005).

[57] Baskin, above n 54, 437.

SCOPE

The third pair of narratives is 'localised <-> globalised' and reflects a more recent tension between seeking to connect climate change to the regional, national, municipal, local, community, household, and indeed, individual[58] level, while also attempting to maintain a global perspective on climate change. Previously, as with many multilateral environmental issues, the 'global' impacted very directly, and usually very causally, upon the 'local'—domestic legislation and other policy action was invariably prompted by what was agreed at the international level. Moreover, it has always been the case that civil society and non-governmental organisations have played a key role in connecting these various spheres of governance.[59] But now, there is a much more complex relationship between domestic policy and national law, on the one hand, and concurrent global negotiations, on the other.

The 2009 session of the Conference of the Parties in Copenhagen, for instance, demonstrated a new—in some ways, altogether different—dynamic than seen either in other multilateral environmental agreements or, even, at previous meetings of the climate change regime itself.[60] For sure, there was much attention on trying to achieve a global deal—and much hand-wringing and disagreement when the best that could be achieved was something significantly less ambitious—but there was also an increased sense, at least amongst some, that effective responses were not solely dependent upon the outcomes of the formal diplomatic process.[61]

[58] See, for instance, P Doran, 'After Copenhagen: Bringing Personal Carbon Trading Home' in D French (ed), *Global Justice and Sustainable Development* (Leiden/Boston, Martinus-Nijhoff, 2010).

[59] See P Spiro, 'Non-Governmental Organizations and Civil Society' in Bodansky, Brunnée and Hey (eds), above n 15, 775–80 who posits that one can identify three strands in liberal theory's account of how NGOs interact within the international environmental law-making process: first, domestic/two-level approaches (where the influence and focus of the NGO is primarily domestic 'mobiliz[ing]...global issues in the same way that they mobilize on domestic ones' (776)); second, transnational approaches (highlighting how 'how national or subnational NGOs work to advance their national, even local, agendas by enlisting foreign and international agents. Key [being]...the development of transnational advocacy networks' (777)); and third, supranational/institutionalist (which very consciously focuses upon the global role of NGOs, though Spiro indicates severe limitations in the current theoretical accounts here as most limit NGO engagement by situating their role and interests 'relative to states' (778)).

[60] See, for instance, J Xercavins i Valls, 'Lessons learnt from Copenhagen: civil society's role in world democratic governance' Global University Network for Innovation (UNESCO/UNU), www.guni-rmies.net/news/detail.php?id=1519 (last accessed: March 2010): 'In the light of events in Copenhagen, what are the paradoxes and lessons learned about civil society participation? That even despite the great transparency and openness of the Conference to the participation of civil society, this participation was reduced to mere observation and protest (which, even if imaginative, was ineffective), with civil society offering very little in terms of an alternative political and propositional discourse regarding what was really happening during the encounter—not to mention before and after'.

[61] Bodansky, above n 17, 10: 'the Copenhagen Accord reflects an apparent shift by China, India, Brazil and South Africa, which begins to break the so-called "firewall" between developed and developing countries. For the first time, the major developing countries agreed to reflect their national emissions reduction pledges in an international instrument... This may seem like a

For instance, at the G-20 Meeting in Pittsburgh in September 2009, though the multilateral process was emphasised—'[w]e will intensify our efforts, in cooperation with other parties, to reach agreement in Copenhagen through the UNFCCC negotiation. An agreement must include mitigation, adaptation, technology, and financing'—this came after, and at best was given only equal weighting to, other initiatives such as the elimination of 'inefficient fossil fuel subsidies' or the acceleration of clean and renewable energy, both of which were viewed as being able to proceed through national action and collaborative effort, without necessarily having a priori endorsement from the climate change Conference of the Parties. As the Pittsburgh Statement noted,

> We call on *all nations to adopt policies* that will phase out such subsidies *world-wide*. [The G20 countries will [t]]ake steps to facilitate the diffusion or transfer of clean energy technology including by conducting joint research and building capacity. The reduction or elimination of barriers to trade and investment in this area are being discussed and should be pursued on a *voluntary basis and in appropriate fora*.[62]

Thus, in a very subtle way, climate change is slowly becoming not only the subject of inter-governmental debate but also a *transnational* one,[63] where domestic regulation and national policy of especially the leading players (both in the developed and developing world) are able to have as much impact upon international behaviour as global negotiation and 'law-making'.[64] For instance, the commitments made by individual countries annexed to the Copenhagen Accord are themselves evidence, if

rather modest achievement; but it represents the first time that these countries have accepted any type of "*internationalization*' *of their national climate change policies*' (emphasis added).

[62] September 24–25 2009, para 31, www.pittsburghsummit.gov/mediacenter/129639.htm.

[63] In addition to this ability of the most influential States to increasingly affect the activity of other States, a related feature is the emergence of 'network administration' and 'hybrid and private administration', on which see B Kingsbury, 'Global Environmental Governance as Administration: Implications for International Law' in Bodansky, Brunnée and Hey (eds), above n 15, 79–80 who highlights the increasingly complex forms of regulation and administration that are now involved in international environmental governance. As he notes, the traditional model (what he terms 'distributed administration'—action 'performed largely by organs of national governments, acting pursuant to international agreements' (64)) is increasingly challenged by the use of other administrative forms, such as networks of (more informal) government officials, private standard-setters and hybrids of both public and private actors. As he notes, 'governance of fundamental global and transnational environmental problems is being displaced from distributed administration, which has been the predominant model… Distributed administration is instead becoming part of the process of diffusion of national environmental law and policy approaches as well as a residuum for special inter-state situations' (ibid).

[64] See, for instance, T Hattori, 'The Role of Domestic Institutions in Constructing a Climate Change Regime: Japan's Formulation of Policy on the Post-Kyoto Framework' (45th ISA Annual Convention, Montreal, Quebec, Canada, March 17–20, 2004), www.allacademic.com//meta/p_mla_apa_research_citation/0/7/2/5/3/pages72533/p72533-1.php: 'International negotiations and domestic institutions are mutually linked in regime formulation processes. The goals of different domestic institutions can lead to different directions in constructing

undoubtedly controversial, of a move away from obligations agreed *within* and *as part of* a global deal, representing the traditional role played by international law in preceding and thus guiding much national action. Indeed, are not such voluntary commitments lacking the necessary normative imprimatur of international law?[65] Without commenting upon the adequacy of the commitments made, the Accord is arguably indicative of a new dynamic between domestic law and global objectives. Whether this is a positive or negative development perhaps only time will tell—both generally and within the context of the climate change regime—but as a potential future model for the transnational coordination of domestic behaviour, it would seem to have a longer-term significance.

Another feature of the broadening scope of the regulatory response to climate change is the increased role of domestic legislators in seeking to forge greater global convergence, without necessarily requiring the endorsement, or explicit approval, of their State's executive machinery. Amongst the plethora of meetings held prior to the Copenhagen Conference of the Parties in 2009, one meeting occurred, which while not overly-significant in and of itself, is nevertheless indicative of the trend highlighted here. Organised by the Global Legislators Organisation for a Balanced Environment (GLOBE),[66] delegations from various countries met to discuss common ways forward. Tens, if not hundreds, of similar meetings—mixing public and private actors—also took place prior to the conference as they do before, and at, many sessions of the parties to multilateral environmental agreements. Nevertheless, the GLOBE meeting stood out as a significant example of the opening up of the diplomatic process not only to non-governmental organisations, which has been well documented, but also to national law-makers. As one of GLOBE's policy objectives for Copenhagen states,

> To be effective, the post-2012 agreement must be underpinned by domestic legislation in all countries. In that context we recognize *the critical role of legislators* in developing and passing legislation, ratification of a post-2012 agreement and in holding governments to account for commitments made.[67]

global regimes, and the role of domestic institutions in the generation of elements of future regimes should not be neglected'.

[65] See UNFCCC, 'Executive Secretary's Clarification Relating to Notification of 18 January 2010' (25 January 2010): 'The phrase "In light of the legal character of the Accord..." should be read in its context. In using the phrase, the secretariat sought to convey two facts regarding the legal nature of the Accord. First, that since the Conference of the Parties neither adopted nor endorsed the Accord, but merely took note of it, its provisions do not have any legal standing within the UNFCCC process even if some Parties decide to associate themselves with it. Secondly, that since the Accord is a political agreement, rather than a treaty instrument subject to signature, a simple letter or note verbale to the secretariat from an appropriate authority in Government is sufficient to communicate the intention of a Party to associate with the Accord'.

[66] See generally, http://copenhagen.globeinternational.org.

[67] copenhagen.globeinternational.org/includes/documents/cm_docs/2009/g/globe_success_criteria_for_cop15_final.pdf, emphasis added.

Of course, one should not make too much of an isolated instance, though this may be one element in a wider debate as to how expansive our understanding should be of what we understand by the term 'climate change regime'. Even if one does not go so far as to indicate that statements such as these indicate a sea-change in approach, the increasing emphasis placed on domestic legislation to contribute to global goals, including 'holding governments to account', is itself a nuanced change worthy of continued investigation and exploration. In part, this reflects a broader debate captured by those who see the evolution in the international system of what they term global administrative law.

> The idea of the emerging global administrative law is animated in part by the view that much of global governance (particularly global regulatory governance) can usefully be analysed as administration. Instead of neatly separated levels of regulation (private, local, national, inter-state) a congeries of different actors and different layers together form a variegated 'global administrative space' that includes international institutions and transnational networks, as well as domestic administrative bodies that operate within international regimes or cause transboundary regulatory effects. The idea of a 'global administrative space' marks a departure from those orthodox understandings of international law in which the international is largely intergovernmental, and there is a reasonably sharp separation of the domestic and the international. In the practice of global governance, transnational networks of rule-generators, interpreters and appliers cause such strict boundaries to break down.[68]

How far global administrative law per se provides a functional analysis of climate change is itself an interesting question,[69] but the notion of 'global administrative space' seems increasingly apposite where domestic (and regional) policies, international governance and non-State actors coalesce around the same—or similar—policy space. More interesting still is the increasing lack of hierarchy between the 'international' and the 'domestic' on climate change matters, which for the international lawyer is not necessarily either the normal or the most comfortable point from which to premise an analysis. Nevertheless, the emerging linkages between the local and the global (as reflected by this pair of narratives) must be recognised as being as significant to the debate as the previous questions over the normativity and the context of the multilateral climate change regime.

[68] B Kingsbury, 'The Concept of "Law" in Global Administrative Law' (2009) 20 *European Journal of International Law* 24.

[69] For a short, but interesting, discussion of the how the perspective of global administrative law can add to the debate, see Kingsbury, above n 63, 75–77 who discusses the administration of the Kyoto Protocol's Clean Development Mechanism, including its Executive Board, noting that 'the result is a highly regulated and bureaucratic structure, whose coherence may suffer where competing political interests pull in different directions... The success of this regime may prove to depend in part of *its inability to monopolize administration*... National or regional regulation may be vital' (77). Emphasis added.

But even if this is the case, how might an international lawyer—bound by the limitations of her/his discipline—usefully contribute to a broadening of climate change beyond the global sphere? Again, a human rights discourse can prove something of an illustration. The overarching—universal—framework that human rights and fairness can provide is likely to prove extremely relevant in ensuring principled and legitimate governance across the regulatory spectrum. In an area with very few normative absolutes, human rights (perhaps less so fairness) will provide the necessary structure to ensure that climate change is tackled not by means of expediency but also in accordance with certain fundamental precepts of international law and community. The very foundation of human rights is the claim that rights protection is a duty first and foremost on national governments. Though the international human rights regime blurs this line by offering protection and *fora* at the global level for addressing concerns—and, indeed, the mainstreaming of human rights into the multilateral climate change debate may contribute still further to the dilution of State responsibilities—it must, however, be remembered that the task of this international interaction and enforcement is ultimately to require States to take responsibility for their actions domestically. To that extent, human rights remind us both of the artificiality of demarcating normative regimes and, more specifically, the overriding responsibilities of States however—and wherever—they choose to undertake regulatory action.

CONCLUSIONS

In drawing to a conclusion, what does all of this tell us about legal understandings of climate change? Ultimately, perhaps little more than climate change as an *object* of study provides a very large—and voluminous—canvas on which to sketch very diverse normative and political suppositions. For instance, one would seem to be able to propose, without being accused of too much inconsistency, both the significance of understanding the current rule- and process-based multilateral system from a doctrinal perspective while synonymously acknowledging the limited nature of that model of governance from a much more critical, or normative, perspective. Or alternatively, that one should be able to focus upon the conventional regime while, at the same time, be able to place it within an increasingly transnational context of domestic and regional norms.

This exploration of ideas should, of course, be positively encouraged, and certainly contributes to our understanding of the issue beyond the paradigm of the multilateral climate change regime *stricto sensu*. As was noted previously, an exclusive focus on the multilateral environmental regime has the potential to ignore the broader framework that is evolving around climate change, elsewhere in international law, in other normative spheres and, more broadly still, in political and socio-economic contexts. But more critically, one might begin to wonder whether this capacity to seemingly accommodate each and every approach—or to use climate change as a conceptual peg for virtually any argument that one might want to construct—is actually problematic, being indicative not only of the scale and complexity of climate change but perhaps, more worryingly, the continued absence of a meaningful consensus as to how best to respond and adapt thereto. Though one cannot dispute the view that 'given the importance of the climate change issue and the slow pace of the international negotiations, there is much to be said for the attitude let a thousand flowers bloom'[70] might it also not be said that such an approach also tells us something about the fundamental conceptual ambivalence at the heart of the climate change debate? And on climate change, is that something we can afford? But that in itself raises a question of the first order; namely, as lawyers, what is our role—as passive observers or meaningful participants? That, however, is an altogether different paper.

[70] D Bodansky, 'Climate Change and Human Rights: Unpacking the Issues', available at http://papers.ssrn.com/abstract=1581555.

Index